'This is an excellent and timely book. It brings together a diverse range of distinguished contributors, who draw on their varied experience in academia, the military, industry, and government. The authors analyse the issues technologically advanced states face as they seek to utilise contemporary military technologies and techniques to maximise military effectiveness in a multi-dimensional battlefield characterised by rapidly advancing technologies and an information-rich environment.'

Michael Sheehan, *Emeritus Professor of International Relations, University of Swansea, UK*

'In this book, the editors have gathered an outstanding set of contributors ranging from scholars to policy professionals to address some of the most pressing technologies (e.g., AI and hypersonics) influencing the rapidly evolving nature of modern war.'

Peter J. Dombrowski, *Professor, U.S. Naval War College*

The Co-evolution of Technology and Warfare

This book explores the relationship between technology and warfare, by examining how recent technological advancements have revolutionized the conduct of war.

The work analyses contemporary conflicts, including the Syrian civil war, the Taliban takeover in Afghanistan, and the ongoing war in Ukraine, but also by exploring future war scenarios and assessing the military capabilities of major powers. In doing so, the book highlights the dynamic and evolving nature of modern warfare. It goes beyond a simple examination of technological advancements, addressing the complexities of modern warfare, scrutinizing the strategies employed by states to adopt and develop military technologies, while emphasizing the importance of technology in shaping military planning, training, research, and innovation. The book provides a collection of timely contributions by leading scholars and practitioners in the military and security field. Furthermore, the contributors identify potential challenges and risks associated with the widespread adoption of technologies in warfare and propose recommendations for policymakers to address issues that relate to military planning and training, research and development, and resilience building.

This book will be of much interest to students of security studies, technology studies, defence studies and International Relations.

Tracey German is a professor of conflict and security in the Defence Studies Department at King's College London, UK. She is the author of *Russia and the Changing Character of Conflict* (2023).

Fotios Moustakis is an associate professor in strategic studies at the University of Plymouth, UK, with over 20 years of experience teaching Royal Navy officers at Britannia Royal Naval College. Since 2015, he has directed the Centre for Sea Power and Strategy (CSS).

Andrew N. Liaropoulos is an associate professor in international relations and strategic studies at the University of Piraeus, Greece. He has taught for years at the Hellenic National Defense College, the Joint Military Intelligence College, the National Security College, and the Air Staff Command College.

Routledge Studies in Conflict, Security and Technology

Series Editors: Mark Lacy, *Lancaster University*, Dan Prince, *Lancaster University*, and Sean Lawson, *University of Utah*

The *Routledge Studies in Conflict, Technology and Security* series aims to publish challenging studies that map the terrain of technology and security from a range of disciplinary perspectives, offering critical perspectives on the issues that concern publics, business and policymakers in a time of rapid and disruptive technological change.

Military Design Thinking
An Historical and Paradigmatic Analysis
Aaron P. Jackson

Theorising Cyber (In)Security
Information, Materiality, and Entropic Security
Noran Shafik Fouad

Creativity in Military Complexity
Design, Disruptors and Defence Forces
Cara Wrigley and Murray Simons

Digital (Dis)Information Operations
Fooling the Five Eyes
Edited by Melissa-Ellen Dowling

Cybersecurity in Latvia
Forging Resilience amidst Emerging Threats
Edited by Mihails Potapovs and Kate E. Kanasta

The Co-evolution of Technology and Warfare
Reshaping the Battlefield
Edited by Tracey German, Fotios Moustakis, and Andrew N. Liaropoulos

For more information about this series, please visit: www.routledge.com/Routledge-Studies-in-Conflict-Security-and-Technology/book-series/CST

The Co-evolution of Technology and Warfare

Reshaping the Battlefield

Edited by Tracey German, Fotios Moustakis, and Andrew N. Liaropoulos

Routledge
Taylor & Francis Group
LONDON AND NEW YORK

First published 2026
by Routledge
4 Park Square, Milton Park, Abingdon, Oxon OX14 4RN

and by Routledge
605 Third Avenue, New York, NY 10158

Routledge is an imprint of the Taylor & Francis Group, an informa business

British Library Cataloguing-in-Publication Data
A catalogue record for this book is available from the British Library

ISBN: 9781032858449 (hbk)
ISBN: 9781032858579 (pbk)
ISBN: 9781003520160 (ebk)

DOI: 10.4324/9781003520160

Typeset in Times New Roman
by Newgen Publishing UK

Contents

Illustrations

Figure

Tables

Contributors

James Henry Bergeron is a seasoned NATO political advisor and an experienced academic who has dedicated 20 years to enabling the success of senior US and Allied leaders and headquarters, successfully advising on strategy and political risk in peace and crisis situations, promoting Allied cohesion, and representing the organisation at diplomatic and political levels. He is an honorary professor of strategic studies at the University of Plymouth, and co-director of eight MARCOM-Plymouth University Sea Power Conferences and a series of high-level seminars on NATO issues. He has 15 years of undergraduate and post-graduate law teaching, research, and programme management experience in Ireland and the UK, including director of post-graduate studies and director of the LL.M. programme at University College Dublin, and director of the Syracuse University Law in London and Politics in England programmes.

Tracey German is a professor of conflict and security in the Defence Studies Department at King's College London. Her research focuses on Russian foreign and security policies, particularly Russia's use of force, as well as Russian strategic culture and military thought. Her latest book, *Russia and the Changing Character of Conflict*, was published by Cambria Press in 2023.

Sidharth Kaushal is a senior research fellow at RUSI. His research covers the impact of technology on maritime doctrine in the 21st century and the role of sea power in a state's grand strategy. Sidharth holds a doctorate in international relations from the London School of Economics, where his research examined the ways in which strategic culture shapes the contours of a nation's grand strategy.

Christopher Lavers is a senior lecturer in Engineering, and has taught maritime littoral and amphibious topics, and earth observation at Britannia Royal Naval College, Dartmouth since 1993. His research focuses on military-to-civilian technologies transfer, preventing the weaponisation of civilian technologies, stealth, and high-resolution imagery for humanitarian crises. He co-managed, delivery outputs for Airbus DS UK, HMG, Exeter, Lincoln, CCW Centre Pembroke Oxford, Plymouth and Southampton, and partners. He has held

30 science-art exhibitions, emphasizing the positive technology applications developed within engineering and physics.

Andrew N. Liaropoulos is an associate professor in International Relations and Strategic Studies at the University of Piraeus. He has taught for years in the Hellenic National Defense College, the Joint Military Intelligence College, the National Security College, and the Air Staff Command College. The period 2015–2021, he has been the director of the laboratory of intelligence and cybersecurity at the University of Piraeus. His research focuses on strategy, intelligence, cybersecurity, hybrid threats, information operations, and new forms of warfare.

Lieutenant Colonel (LtCol) Daniel Love joined the George C. Marshall European Center for Security Studies in June 2023 as a military professor and, as of February 2025, has served over 17 years in the United States Marine Corps as an officer, a foreign security forces advisor to the Afghan National Army, a Eurasia foreign area officer with Ukrainian language coding, and a Northeast Asia regional affairs officer. LtCol Love holds dual Masters of Arts (MA) from University of California San Diego (MA in Security of the Asia-Pacific) and Naval Postgraduate School (MA in National Security Studies of Europe and Eurasia), and has Ukrainian language professional proficiency.

Fotios Moustakis is an associate professor in strategic studies at the University of Plymouth, with over 20 years of experience teaching Royal Navy Officers at Britannia Royal Naval College, Special Forces, international officers, and university students in strategy, security, and counterterrorism. He has served as a visiting professor at NATO's Defence Against Terrorism course and was a senior associate member at St Antony's College, Oxford. His research focuses on strategy, security, and international terrorism, with publications in academic and policy journals. Since 2015, he has directed the Centre for Sea Power and Strategy (CSS), shaping security policy and fostering education, research, and collaboration. In 2009, he pioneered the first blended learning International MA in applied strategy and international security for senior staff at the Hellenic National Defence College, Greece.

Matthew Mowthorpe currently works in the UK Military Future Programmes at Airbus Defence and Space. Prior to this, he worked at the Ministry of Defence, where he managed the Space Team examining threats to and from Space. Dr Mowthorpe has published in numerous journals on the weaponization of space and notably published the book *The Militarization and Weaponization of Space* published by Rowman and Little in the US.

Jack Sharpe is a major in His Majesty's Royal Marines and a PhD candidate in politics at the University of Plymouth. His thesis focuses on the impact of space and cyber domains on maritime security. More broadly, he is a cyber security and technology executive focused on strategic and change leadership, organisational transformation, and business development.

Markos Trichas is the director of National Security and Defence Space within BAE Systems. Prior of joining BAE in 2023, he worked at Airbus Defence and Space, Harvard-Smithsonian Centre for Astrophysics, Rutherford Appleton Laboratory, and Imperial College London. He holds a PhD in astrophysics from Imperial College London. He is also a visiting professor at Plymouth University and has authored numerous peer-reviewed publications.

Konstantinos Tsetsos is the head of foresight at the Metis Institute for Foresight and Strategy at the Universität der Bundeswehr München, Germany. His work focuses on international security, war theories, crisis management, political risks, and future analysis.

Foreword

This edited volume brings together 11 leading subject matter experts (military practitioners, analysts, and academics) to explore how rapid advances in transformative technologies revolutionize warfare, constantly reshaping the scale, scope, duration, and dynamics of conflict. This process intensifies daily and appears bewildering and difficult to grasp. To give an example, Ukraine experiences what is now the largest war in Europe since 1945, and in late March 2025 set itself the goal of manufacturing 15,000 military use robots in 2025 (logistics, mining, and combat). Simultaneously, the Netherlands announced a €500 million support for Ukraine's attack drone development program and, the commander of the Ukrainian armed forces stated that Ukraine had destroyed 408 Russian army artillery systems in the last week of March (4005 artillery systems in the first three months of 2025), which Russia compensates by increasing its use of attack drones and glide bombs.

This book identifies how a range of rapidly advancing transformative technologies, not least Artificial Intelligence (AI), robotics, cyber warfare, and autonomous systems, are used in warfare. It highlights the associated risks, challenges and opportunities this interaction generates, as well as assess the broader implications and outcomes. Through 10 chapters, *inter alia*, the expert contributors address: the interaction of AI and cyberwarfare; AI and information operations; how state and non-state actors utilize unmanned aerial vehicles (UAVs); space warfare and Russian and Chinese counter-space warfare programs; the impact of Russian hypersonic weapons in reshaping conflict calculus; how globalization processes reshape the structure of global shipbuilding and impact sea power; the emergence of a second revolution in military affairs (2RMA), not least in the maritime domain; cyber threats to critical infrastructure and the implications for resilience, collaboration and public awareness; and how Ukraine's unmanned aerial systems (UAS) and artillery operations and counter-UAS are critical to success in 21st-century warfare.

The implications of the evolving nexus between technologies and the pace of revolution in warfare are profound. As such, this book will be of great interest to policy makers, welcoming of the promises of strategic foresight but sensitive to how military superiority can be accompanied by vulnerability and unforeseen consequences. Military practitioners interested in leadership, force structure, training, doctrine, tactics, and strategy will find this book valuable, as will defence industry focussed on production at scale and cost. Finally, the book appeals to analysts and academics grappling with complexities generated by the interconnected and inter-enabling relationship between technology and war, as well as the implications themselves for the conduct of war, the necessity for state institutional adaptation, the distribution of military power in the international system, and the structure of that system itself.

Dr. Graeme P. Herd
Research and Policy Analysis Department
George C. Marshall European Center for Security Studies (GCMC)

Acronyms and Abbreviations

A2AD	Anti-access/Area Denial
ABC	Actor, Behaviour, Content
ABL	Airborne Laser
ADRV	Advanced Debris Removal Vehicle
AFU	Armed Forces of Ukraine
AI	Artificial Intelligence
AIS	Automatic Identification System
AKM	Apogee Kick Motor
AL	Aolong
ASAT	Anti Satellite
ASCMs	Anti-Ship Cruise Missiles
ASEAN	Association of Southeast Asian Nations
ASW	Anti-Submarine Warfare
AT	Anti-Tank
ATC	Air Traffic Control
BMD	Ballistic Missile Defence
C2	Command and Control
C2I	Command and control information
CBRN	Chemical Biological, Radiological or Nuclear
CCTV	Closed-circuit television
CI	Critical Infrastructure
CNN	Cable News Network
COP	Common Operating Picture
COTS	Commercially-Off-The-Shelf
CSIS	Center for Strategic and International Studies
CUI	critical undersea infrastructure
DARPA	Defense Advanced Research Projects Agency
DDoS	Distributed Denial-of-Service
DIA	Defence Intelligence Agency
DIB	Defense Industrial Base
DJI	Da-Jiang Innovations
DL	Deep Learning

DLP	Data Loss Prevention
DN	Dong Neng
DoD	Department of Defense
EMP	Electromagnetic Pulse
EW	Electronic Warfare
FAA	Federal Aviation Administration
FOV	Field Of View
FPV	First Person View
GANs	Generative Adversarial Networks
GDP	Gross Domestic Product
GDPR	General Data Protection Regulation
GEO	Geostationary Orbit
GNSS	Global Navigation Satellite System
GPS	Global Positioning System
GSC	Government Security Classification
GTOW	Gross Take-off Weight
HAWC	Hypersonic Air-breathing Weapon Concept
HCM	Hypersonic Cruise Missile
HGV	Hypersonic Glide Vehicle
ICBM	Intercontinental Ballistic Missile
ICTs	Information and Communication Technologies
IDF	Israel Defense Forces
IED	Improvised Explosive Device
IOs	Influence Operations
IPS	Intrusion Prevention Systems
IRA	Internet Research Agency
ISR	Intelligence, Surveillance, and Reconnaissance
IT	Information Technology
LEO	Low Earth Orbit
LLMs	Large Language Models
LSCO	Large-Scale Combat Operations
MAD	Mutually Assured Destruction
MEO	Medium Earth Orbit
ML	Machine Learning
MOD	Ministry of Defence
NASA	National Aeronautics Space Administration
NATO	North Atlantic Treaty Organisation
NAVAIR	Naval Air Systems Command
NAVPLAN	Navigation Plan
NGO	Non-Governmental Organization
NLP	Natural Language Processing
NRO	National Reconnaissance Organisation
OAE	Operation Active Endeavour
OSCE	Organization for Security and Co-operation in Europe
OT	Operational Technology

PGM	Precision Guided Munition
PLA	People's Liberation Army
PLAN	People's Liberation Army Navy
PNT	Positioning, Navigation and Timing
PRC	People's Republic of China
R&D	Research and Development
RCS	Radar Cross Section
RF	Radio Frequency
RF	Russian Federation
RFN	Russian Federation Navy
RMA	Revolution in Military Affairs
RPG	Rocket Propelled Grenade
RPO	Rendezvous Proximity Operations
RTO	Return-To-Origin
SAR	Synthetic Aperture Radar
SIEM	Security Information and Event Management
SJ	Shijian
SLOCs	Sea Lines of Communication
SME	Small and Medium-Sized Firm
SOC	Security Operations Centers
SSGN	Guided-Missile Submarines
SY	Shiyan
TCG	Turkish Republic Ship
TEL	Transport Erector Launch
TJS	Tongxin Jishu Shiyan
TS	Tansuo
UAS	Unmanned Aerial Systems
UAV	Unmanned Aerial Vehicle
UCAV	Unmanned Combat Aerial Vehicle
UK	United Kingdom
US	United States
USN	United States Navy
USSR	Union of Soviet Socialist Republics
USVs	Undersea Vehicles
UtDs	UAV-to-drones
VAEs	Variational Autoencoders
VUCA	Volatile, Uncertain, Complex, and Ambiguous
WMD	Weapons of Mass Destruction
WWII	World War Two
WWI	World War One

1 Introduction

Tracey German, Fotios Moustakis, and Andrew N. Liaropoulos

Debating the future of war

Throughout history, warfare has been shaped by geopolitical, societal, economic, military, and technological factors. These same factors continue to influence contemporary debates on the future of warfare, making it one of the most contentious topics among academics, militaries, and policymakers. Assertions about the apparent transformation of conflict are not new; what is new is the pace of such change, accelerated by the ongoing technological and communications revolution. The spread and intensification of violence over the past two decades has demonstrated once more the evolving character of war from hybrid and surrogate conflicts to unrestricted and postmodern warfare. War in the 21st century has become increasingly non-linear and unpredictable, involving a complex array of actors, both state and non-state, who use a variety of military and non-military means to influence and coerce an adversary. Rapid changes in technology and communications, as well as the growing interconnectedness of societies, has added to the complexity.

Recent developments in military technology and weapon systems have sought to enhance the effectiveness of kinetic force. However, evidence from recent conflicts, including the Syrian civil war, the Israel–Hamas war in the Gaza Strip, and Russia's ongoing war in Ukraine, demonstrates the limitations of military innovations, especially in the case of prolonged conflicts and the emergence of unconventional actors and warfighting strategies. Moreover, the rapid evolution of information technology, the proliferation of global communication networks, and the weaponisation of artificial intelligence (AI) have added further complexity to contemporary conflicts. These developments highlight the unpredictability surrounding the future of warfare. Speculating on the future of warfare is truly a difficult task. Thinking about the next war is usually limited by attempts to extrapolate from current trends in order to speculate on future developments. It is common to think linearly about what might happen, based on what has happened. Consequently, the experiences of individual states foster different visions of future conflict. This book does not attempt to predict the future of war. Instead, it provides a critical analysis of how recent technological advancements are reshaping military strategies, tactics, and the broader conduct of warfare.

DOI: 10.4324/9781003520160-1

One of the most striking characteristics of contemporary warfare is its increasing complexity. Modern conflicts are often fought across multiple domains, including land, sea, air, space, and cyberspace. The distinction between combatants and non-combatants has also become more ambiguous, as adversaries employ unconventional tactics such as cyber-attacks, disinformation campaigns, and proxy warfare to achieve strategic objectives. Furthermore, the manipulation of interconnected and information-rich environments makes it increasingly difficult to differentiate between friend and foe. Advanced information and communications technologies enable subversive efforts, with the internet and social media platforms amplifying divisive narratives and disinformation, serving as tools of indirect action that target a society's ability to resist. Moreover, globalisation has played a significant role in shaping the dynamics of warfare. The interconnected nature of the global economy, supply chains, and communication networks means that localised conflicts can have far-reaching implications. Economic warfare, sanctions, and the disruption of critical infrastructure are now integral components of modern strategy. In an era where technological advancements are accelerating at an unprecedented pace, understanding how these shifts influence the character of war is crucial for policymakers, military planners, and academics alike.

Potential changes to the character of conflict have real-world implications: how states envision the character of conflict shapes how they plan and prepare for war, from defence policy to procurement, and from doctrine to training. As warfare continues to evolve, so too must the doctrines and strategies that guide military operations. The ability to adapt to new forms of conflict will determine the success or failure of future military engagements. With this in mind, this book examines not only the technological aspects of modern warfare but also the broader implications for military strategy, national security, and international stability.

The co-evolution of technology and warfare

History demonstrates that the relationship between technology and warfare is complex and dynamic. As new technologies emerge, they redefine military strategies and tactics. On the other hand, warfare drives technological innovation, often producing advancements beyond the conventional battlefield. The term 'military technical revolution', coined by the Soviets in the 1980s, was a precursor to the idea of a 'revolution in military affairs' (RMA), which gained prominence in the US in the 1990s, arguing that technology on its own is insufficient to drive major military change; RMAs require operational innovation and changes in doctrine and organisation, alongside new technology. The end of the Cold War and the 1991 Gulf War renewed the debate about RMAs and the impact of new technologies on the conduct of war, with disagreement about both definition and occurrence.[1]

The invention of gunpowder revolutionised warfare by making fortifications obsolete and leading to modern firearms. The Second World War led to advancements in radar systems, jet engines, and computing. Technologies designed for war often have civilian applications (e.g. global positioning systems, drones). Today, the co-evolution of technology and warfare is being shaped by the Fourth

Industrial Revolution – a transformation distinct from but built upon the Digital Revolution. The Fourth Industrial Revolution blurs the distinction between physical, digital, and biological domains. Emerging technology breakthroughs in areas as diverse as AI, robotics, the Internet of Things (IoT), three-dimensional printing, nanotechnology, and quantum computing profoundly affect every aspect of our society. As with previous technological revolutions, these advancements are reshaping societies while simultaneously transforming the character of warfare. Modern militaries are being forced to rethink how they train, organise, and conduct operations in response to these disruptive changes.

Indicative of the above discussion is the application of AI in the military domain. The applications range from image recognition for surveillance to situational awareness on the battlefield and from logistics for battle management to a compressed decision-making loop. AI should not be understood as a weapon, but rather as an enabler and force multiplier of a wide range of military capabilities, both kinetic and non-kinetic. One potential application of AI is the use of autonomous weapons systems (AWS). Algorithms program such weapons systems to select and engage targets based on predetermined criteria; thus, they can operate without direct human intervention. Such systems reduce risk to human lives while enhancing operational capabilities, allowing for continuous surveillance and rapid deployment. The proliferation of AWS has shifted the dynamics of warfare, enabling asymmetrical warfare where smaller forces can effectively challenge larger, conventionally superior adversaries.

Another example that demonstrates the vast impact of technology on warfare is that of cyberspace. Cyberoperations targeting critical infrastructure, communication networks, and command systems can disrupt entire nations without the need for traditional kinetic force. Such capabilities allow states and non-state actors to wage covert operations and undermine enemy defences. Adding to that, cyberspace offers anonymity and plausible deniability. In some cases, these characteristics make cyberoperations the preferred choice. Furthermore, advancements in biotechnology and nanotechnology could significantly impact future battlefields. Biometric sensors, exoskeletons, and cognitive augmentation have the potential to enhance human performance and push the boundaries of warfare. Nanotechnology-based material can improve body armour or create adaptive camouflage that blends with surroundings or even absorb radar signals, thus making soldiers and their weapons systems more agile and stealth.

Another frontier in modern warfare is the militarisation of outer space. Satellites are integral to military operations (and many facets of daily life in modern societies) providing navigation, communication, surveillance, and reconnaissance capabilities. However, these systems are increasingly vulnerable to anti-satellite weapons and jamming tactics that could disrupt not only military functions, but also global connectivity and thereby global supply chains. As strategic competition in space intensifies, the need for international agreements and norms to prevent conflict escalation has never been more critical.

These developments illustrate the profound ways in which warfare is being redefined by the rapid integration of emerging technologies. At the same time, while technological advancements present new opportunities for military superiority, they also introduce vulnerabilities and unforeseen consequences. The increased reliance on AI-driven decision-making, automated warfare, and cyber capabilities raises questions about accountability, escalation risks, and the potential for unintended conflicts. The challenge for policymakers and military planners is to balance innovation with ethical responsibility, ensuring that technological progress does not outpace strategic wisdom.

Another pressing concern is the potential for an arms race in emerging technologies. As nations compete to develop more advanced weaponry, the risk of destabilisation grows. The lack of clear international regulations governing AI weapons, cyberwarfare, and space militarisation increases the chances of miscalculations and escalation.

Furthermore, as new technologies become integrated into modern warfare, there is a growing need for interdisciplinary expertise. Military strategists must collaborate with technologists, ethicists, and policymakers to develop frameworks that ensure emerging innovations serve not only military efficiency but also broader security and stability. Adaptability and strategic foresight will be critical in managing the rapid evolution of warfare, ensuring that technological advancements are leveraged responsibly while minimising the risks they pose.

By exploring the intricate relationship between warfare and technology, this book aims to provide a comprehensive and critical analysis of how technological progress is reshaping military strategies, operational doctrines, and the geopolitical landscape. The coming decades will witness profound shifts in the art of war, and understanding these changes is imperative for those shaping the future of global security.

Overview and main themes of the book

This book delves into the intricate relationship between technology and warfare and explores how recent advancements in technology have revolutionised the conduct of war and the consequences they carry for international security. Warfare has always been a dynamic arena where technology plays a pivotal role in shaping strategies, tactics, and outcomes. One of the enduring features of conflict over the centuries has been its state of flux: echoing Carl von Clausewitz, war is a true chameleon. The character of war in the 21st century continues to transform, adopting different forms, including cyberwarfare and cognitive warfare. Highly developed armies use both kinetic and non-kinetic means, fighting with missiles and algorithms, and weaponising robots and strategic narratives. This book provides policymakers, military strategists, and researchers with critical insights into the changing dynamics of modern warfare. It addresses the complexities of modern warfare, by stressing the key role of technology and examining strategies that states employ to adopt and develop technologies for military purposes. Furthermore, this book identifies potential challenges and risks associated with the widespread adoption of technologies in warfare and proposes recommendations for policymakers to address issues that relate to military planning and training, research and development, and resilience building. The contributors are leading experts in their field, and combine theoretical knowledge and practical experience from the academia, the industry, the military, and the government. Their collective expertise enables this book to effectively bridge the divide between theory and practice.

This book goes beyond a simple examination of technological advancements, addressing the complexities of modern warfare. It scrutinises the strategies employed by states to adopt and develop military technologies, emphasising the importance of technology in shaping military planning, training, research, and innovation. Such an approach has been lacking in the relevant literature and, thus, the book aims to fill this gap and contribute to the discussion on the future of warfare.

In terms of structure, the book consists of twelve chapters. The introduction is followed by ten chapters that cover a wide range of issues relating to how technology affects the conduct of warfare. Finally, the concluding chapter summarises the key findings, connecting the dots regarding the co-evolution of technology and warfare, and explores areas for further research.

Jack Sharpe presents his viewpoint on the intersection of AI and cyberwarfare in Chapter 2. He examines the advantages and disadvantages associated with the use of AI in cyber operations. The conflict in Ukraine serves as a testing ground for AI-enhanced cyberwarfare, which Sharpe explores, in order to address the legal and ethical dilemmas of this new form of warfare. The chapter ends with a set of recommendations for policymakers, military leaders, and cybersecurity professionals to balance innovation with ethical responsibility.

In Chapter 3, Andrew Liaropoulos discusses how advancements in AI and in particular in generative models have shaped the conduct of influence operations (IOs). He documents how AI-powered tools have significantly transformed the ability to create and disseminate dynamic, personalised, and real-time content to targeted audiences. Chatbots, deepfake technologies, and microtargeting are the new weapons in the battle over perceptions and attention. The chapter not only provides real-world examples where AI-powered tools have been integrated in the conduct of IOs but also highlights future trends and limitations.

In Chapter 4, Christopher Lavers offers a holistic overview of the role of unmanned aerial vehicles (UAVs) and stresses the strategic and tactical advantages that modified drones provide to urban guerrillas and terrorist groups. The analysis also considers the technological vulnerabilities of UAVs, including cyber threats, and countermeasures, before concluding with the broader implications of drone proliferation. Ultimately, the analysis underscores the urgent need for policy interventions and technical safeguards to prevent the weaponisation of consumer UAV technology.

The role of space warfare and the counter-space programs by Russia and China are presented in Chapter 5. Markos Trichas and Matthew Mowthorpe examine China and Russia's ASAT (Anti Satelite) concepts in the context of their respective military space doctrines. China recently demonstrated a new capability to hide in the 'graveyard' beyond geostationary orbit (GEO) and re-emerge to grapple a satellite in GEO. Russia in 2024 demonstrated the intent to place nuclear weapons in space shattering the Outer Space Treaty's prohibition of placing weapons of mass destruction in space. The authors conclude that space is a contested domain, necessitating innovative defence strategies.

In Chapter 6, Tracey German explores the role that hypersonic weapons could play in reshaping the dynamics of conflict. These weapons have not only raised questions over deterrence but have also challenged traditional notions of defence and offence. The chapter sets out the key characteristics of hypersonic missiles and the challenges presented by such weapons, and explores the Russian example to demonstrate why states decide to pursue hypersonic technology, as well as the potential pitfalls.

Sidharth Kaushal examines how globalisation has reshaped naval strategy, particularly in the context of the US–China maritime rivalry in Chapter 7. He argues that sea control is shifting from distant blockades to operations in littoral regions. The complexity of modern shipping – frequent flag changes, fragmented ownership, and at-sea transactions – makes traditional interdiction more difficult. Such developments favour China, which, with its geographic proximity and expansive coast guard, can exert greater control over regional trade than the US. However, the US is adapting by leveraging strategic advantages such as advanced submarines and long-range missile capabilities. Thus, shifts in the structure of the global shipbuilding industry raise questions regarding where the locus of 21st-century sea power (as opposed to naval power alone) will be.

In Chapter 8, James Henry Bergeron demonstrates that previously held assumptions about maritime strategy in the post-Cold War era have collapsed. Emerging naval powers and advanced technologies mark a Second Revolution in Military Affairs (2RMA), exposing NATO's aging fleets and logistical struggles. According to James Henry Bergeron, a revised maritime strategy must align with evolving geopolitical realities and technological innovations, emphasising economic strength, innovation speed, and cultural reassertion of sea power.

The extent to which hybrid and cyber threats pose a challenge to critical infrastructure is explored in Chapter 9. Konstantinos Tsetsos analyses the nature of such threats and outlines key recommendations that enhance resilience. Strengthening cybersecurity, fostering international collaboration, enhancing public awareness, and integrating civilian and governmental resources are central to addressing these threats.

In Chapter 10, Fotios Moustakis delves into the evolving role of military leadership and training in the context of 21st-century high-tech warfare, characterised by the rapid advancement of transformative technologies such as AI, robotics, cyber warfare, and autonomous systems. The chapter stresses the critical importance of transformational leadership and highlights the importance of cultural awareness, continuous education, and the human dimension of leadership in navigating modern security challenges.

Chapter 11 examines Russia's war in Ukraine. Daniel Love focuses on the employment of unmanned aerial systems (UAS) and artillery operations by the Armed Forces of Ukraine (AFU) against the Russian forces. He concludes that

UAS, counter-UAS, and artillery operations are crucial aspects of this war and will be critical to success in 21st-century warfare.

To conclude, in the above chapters, we have secured the participation of a very pluralistic and expert group of authors from different backgrounds. Their approaches and views are central to a book that aims to make a useful contribution to the current debate on how the Fourth Industrial Revolution is already affecting the conduct of warfare.

Note

1 See Martin Van Creveld, *Technology and war: From 2000 BC to the present* (London: Simon and Schuster, 2010); Steven Metz, *Strategy and the revolution in military affairs: From theory to policy* (Collingdale, PA: Diane Publishing, 1995); EC Sloan, *Revolution in Military Affairs* (McGill-Queen's Press-MQUP, 2002).

2 Artificial Intelligence and Cyber Warfare

The New Battleground

Jack Sharpe

Introduction

It is essential to note that artificial intelligence (AI) has myriad uses outside of the defence and strategic influence domains. Nevertheless, the convergence of AI and cyber warfare has advanced digital conflict, fundamentally reshaping the landscape of global security. Cyber warfare, broadly defined as the use of cyber-attacks against an enemy state to inflict damage comparable to actual warfare or disrupt vital computer systems, has evolved significantly over the past decades (Fortinet *n.d.*). This form of conflict encompasses a range of actions aimed at espionage, sabotage, propaganda, manipulation, and economic warfare. While ongoing debates among experts persist regarding the precise definition of cyber warfare, it involves state or state-sponsored actors targeting another nation's digital infrastructure (Heitzenrater 2023). Crucially, AI can enable decision advantage, efficiency, new capabilities, and whole-force empowerment (MoD 2022), thereby acting as a significant force multiplier in the delivery of cyber warfare.

According to The Alan Turing Institute (*n.d.*), AI is a technology ecosystem that broadly includes:

- AI (Hardware): Machines that perform tasks that would previously have required human (or other biological) brainpower to accomplish. This is a broad field that incorporates many distinct aspects of intelligence, such as reasoning, spatial awareness, and problem-solving.
- Machine Learning: A subset of AI for computer algorithms that can 'learn' by finding patterns in sample data and then apply the observations to produce useful outputs or predictions, often using neural networks.
- Data Science: Research that involves the processing of substantial amounts of data to provide insights into real-world problems.

The degree of AI ranges from manual or human operation, through automation, towards autonomy. Each stage becomes increasingly less dependent on human input or assistance until no longer requiring human input.

DOI: 10.4324/9781003520160-2

The scope of cyber warfare has expanded to encompass various tactics within the non-kinetic effects arsenal. This expansion reflects a transformation in the character of warfare towards higher lethality supported by non-physical, high-tech domains. For context, this chapter will specifically focus on the employment of AI and cyber capabilities during military operations, a specific application that is subordinate to their broader utility in statecraft. The integration of cyber and AI technologies has significantly advanced how competitive and strategic advantages are achieved, paving the way for the efficacious delivery of sub-threshold and hybrid warfare (Sharpe *et al.* 2024; Bachman 2023; Petersen 2023; Cullen & Reichborn-Kjennerud 2017). The primary focus areas of these AI and cyber warfare activities include:

- Attacks on critical infrastructure systems.
- Espionage and theft of sensitive information.
- Denial-of-service attacks to disrupt essential services.
- Propaganda and disinformation campaigns.
- Attempts to compromise power grids and communication networks.

As nations continue to develop their offensive and defensive cyber capabilities, the potential for physical confrontation resulting from cyber operations has increased. Notable examples of cyber warfare incidents include the 2015 Chinese hack of the U.S. Office of Personnel Management and the use of NotPetya ransomware against Ukraine in 2017. More recently, the ongoing Russia–Ukraine conflict has underscored the role of cyber warfare in modern geopolitical disputes, with attacks targeting Ukrainian organisations and infrastructure (Lewis 2022). Initially focused on straightforward denial-of-service attacks and data breaches, cyber warfare now encompasses a wide range of sophisticated strategies and technologies across a plethora of activities.

AI plays a crucial role in cybersecurity by enhancing the speed, scale, and sophistication of both offensive and defensive cyber operations (Imperva *n.d.*; IBM *n.d.*). AI-powered systems can swiftly identify and respond to security incidents, accelerate threat detection and mitigation, and optimise analysts' time in managing complex cyber threats (IBM *n.d.*). Furthermore, AI technologies enable the automation of complex tasks, such as vulnerability detection, data analysis, and decision-making, which were traditionally performed by human operators (Amster 2024; Palo Alto 2024a). This automation not only accelerates the pace of cyber-attacks but also increases their effectiveness and unpredictability. For instance, AI-driven malware can learn from its environment and evolve to bypass security measures (Fitzgerald & Bonnie 2024), while AI-powered disinformation campaigns can manipulate public opinion with unprecedented precision (IBM *n.d.*).

Unsurprisingly, the intersection of AI and cyber warfare has become a critical focus in contemporary global security discourse. The rapid advancement of AI technologies has significantly enhanced the capabilities of cyber operations, presenting novel challenges to international stability and security (Lindsay 2020).

Major powers, particularly Russia and China, have made substantial investments in leveraging AI to augment their cyber capabilities, potentially shifting the balance of power in the digital domain (Horowitz *et al.* 2018). These developments raise profound questions about the nature of future warfare, the effectiveness of traditional defence mechanisms, and the ethical implications of autonomous cyber weapons (Scharre 2018). The integration of AI into cyber operations not only accelerates the pace and scale of attacks but also introduces new levels of complexity and unpredictability to threat landscapes (Brundage *et al.* 2018). This evolution in cyber capabilities necessitates a re-evaluation of existing security paradigms and the development of new strategies to address emerging threats. The potential for AI-enhanced cyber warfare to disrupt critical infrastructure, manipulate information ecosystems, and even interfere with critical command and control systems underscores the urgency of addressing this issue at both national and international levels (Geist & Lohn, 2018). As the line between cyber espionage and offensive operations becomes increasingly blurred, the risk of unintended escalation in conflicts also grows (Buchanan 2020).

AI-Driven Strategies: Risks and Advantages

The integration of AI into cyber warfare has significant strategic advantages and considerable risks. This dual nature of AI in cyber operations necessitates a deep understanding of its implications upon the requirement to develop effective cybersecurity strategies and policies as well as militarily defensive and offensive cyber capabilities (Brundage *et al.* 2018).

One of the most notable advantages of AI in cyber warfare is its ability to exponentially enhance the speed and scale of attacks. Machine learning algorithms automate the identification of targets and execution of attacks, enabling unprecedented rapidity and volume. For instance, AI-powered botnets can launch distributed denial-of-service (DDoS) attacks at scales and speeds that far surpass traditional methods (Taddeo *et al.* 2020). Furthermore, automated malware can propagate and adapt to new environments with alarming speed, potentially infecting vast networks in minimal time.

AI also improves the ability to detect and exploit vulnerabilities. These systems excel at pattern recognition and can be trained to identify weaknesses within systems more efficiently than human analysts (Apruzzese *et al.* 2018). By continuously scanning for vulnerabilities, AI can often identify zero-day exploits before they are patched, thereby enhancing both offensive and defensive cyber operations. Additionally, machine learning models can predict potential vulnerabilities in new software releases, allowing for pre-emptive exploitation or defence.

In intelligence gathering, the capacity of AI to process and analyse large datasets provides a substantial advantage. Natural language processing (NLP) algorithms can sift through vast amounts of data from various sources, identifying relevant information for intelligence purposes with remarkable efficiency (Kott *et al.* 2015). This capability enables cyber operators to gather insights about potential targets, thereby enhancing the effectiveness of their operations. The ability of AI systems to make real-time decisions during cyber operations significantly enhances their effectiveness and responsiveness. These systems can autonomously adjust attack vectors based on the target's defences, maximising the chances of success (Horowitz *et al.* 2018). Machine learning algorithms can prioritise targets and allocate resources in real-time, optimising the impact of cyber operations and allowing for a level of adaptability previously unattainable (Sontan & Samuel 2024).

However, the advantages of AI in cyber warfare come with significant risks. The speed and autonomy of AI-driven cyber-attacks could lead to rapid escalation of conflicts. It is conceivable that AI systems could misinterpret situations or overreact to perceived threats, potentially triggering unintended conflicts (Geist & Lohn 2018). The rapidity of these attacks could leave little time for human intervention or diplomatic resolution, exacerbating tense geopolitical situations.

The complexity of AI systems also introduces new levels of unpredictability in cyber operations. AI models may behave in unexpected ways when faced with novel situations, potentially causing unintended consequences (Johnson 2019). The 'black box' nature of some AI algorithms makes it difficult to fully understand and predict their decision-making processes, adding an element of uncertainty to cyber operations.

Attribution of attacks becomes increasingly challenging with AI-driven cyber operations. These attacks can be designed to obscure their origin, with AI generating false flags and misleading indicators that complicate efforts to identify the true source (Buchanan 2020). Similarly, autonomous systems can create plausible deniability for state actors and their proxies engaging in cyber operations, further muddying the waters of international cyber conflicts and jeopardising broader geopolitical stability.

Lastly, the development and deployment of autonomous AI-driven cyber weapons raise significant ethical questions. Concerns about the lack of human oversight in potentially destructive cyber operations are paramount (Etzioni & Etzioni 2017). The use of AI in cyber weapons challenges existing legal frameworks and international norms regarding warfare, necessitating a re-evaluation of the West's ethical standards in this new digital battleground.

While AI offers substantial advantages in cyber warfare, it also introduces new risks and ethical challenges. The task ahead lies in balancing these factors to develop responsible and effective AI-driven cyber strategies. As Western powers continue to navigate the complexities of integrating AI across political, military, and business endeavours, it is imperative that defence and security experts remain attuned to the opportunities and dangers presented in cyber warfare (Lindsay 2020). When comparing the UK with other major powers, such as the United States, China, and Russia, it is apparent that while all these nations recognise the strategic importance of cyber capabilities, their approaches and resources vary significantly (Lindsay *et al.* 2015). The UK, for instance, has developed a robust cybersecurity framework, but it faces challenges in keeping pace with the rapid advancements made by its counterparts (Dewar 2018).

Emerging Challenges in Cyber Infrastructure Security

The fast-paced amorphous integration of AI and cyber operations presents a new set of challenges for securing cyber infrastructure, reshaping the cybersecurity landscape, and necessitating innovative approaches to defence. One of the most significant challenges posed by AI in cyber warfare is its ability to continuously evolve and outpace traditional defence mechanisms. AI-powered malware, for instance, can adapt to its environment, learning from failed attempts and adjusting its tactics to bypass security measures. This adaptability makes such threats particularly difficult to detect and mitigate using conventional methods (Brundage *et al.* 2018).

Attackers are also developing AI models specifically designed to deceive other AI systems, potentially compromising AI-based security tools. These adversarial attacks can exploit vulnerabilities in machine learning models, rendering them ineffective or causing them to make incorrect decisions (Goodfellow *et al.* 2015). Furthermore, AI systems can quickly identify and exploit newly discovered vulnerabilities, often faster than human defenders can patch them. This speed advantage can lead to large-scale attacks before adequate defences can be implemented (Geist & Lohn 2018).

The inherent complexity of AI systems introduces new challenges in ensuring their security and integrity. Many AI models are opaque and operate as 'black boxes' making it difficult to infer their decision-making processes (Palo Alto 2024b). This lack of transparency can complicate efforts to identify and address security vulnerabilities (Rudin 2019). Additionally, as AI systems often rely on pre-trained models and datasets from various sources, ensuring the integrity of these components becomes crucial. Compromised or maliciously altered training data or models could introduce backdoors or biases into AI-based security systems (Steinhardt *et al.* 2017). Protecting proprietary AI models[1] from theft or reverse engineering is also challenging, potentially allowing adversaries to develop targeted attacks or countermeasures (Tramèr *et al.* 2016).

Of note, AI is enhancing the effectiveness and scale of social engineering attacks as AI-generated audio and video can create highly convincing impersonations, potentially fooling even careful individuals and bypassing traditional authentication methods (Chesney & Citron 2019). AI can analyse vast amounts of personal data to create deliberately targeted and convincing phishing attacks, making them much harder to detect. Moreover, AI-driven bots can engage in large-scale social media manipulation, spreading disinformation or influencing public opinion with unprecedented efficiency (Ferrara *et al.* 2016).

The dynamic nature of AI-powered threats necessitates ongoing improvements in AI defence mechanisms. AI-based defence systems require continuous retraining with up-to-date data to remain effective against evolving threats. This process is resource-intensive and challenging to maintain in real-time environments (Apruzzese *et al.* 2018). Tuning AI defence systems to minimise false positives while maintaining high detection rates is an ongoing challenge, requiring constant refinement. As attackers develop new AI-powered techniques, defenders must continuously update their AI models to recognise and counter these emerging threats

(Buchanan 2020). Furthermore, AI systems must be capable of defending against threats across various platforms and domains, requiring extensive training on diverse datasets and scenarios (Hallaq *et al.* 2017).

These emerging challenges in cyber infrastructure security underscore the need for a revolutionary shift in cybersecurity approaches. Organisations and governments must invest in research, development, and implementation of AI-driven security solutions while also addressing the ethical and practical challenges they present. Collaboration between academia, industry, and government agencies will be crucial in developing robust, adaptable, and ethical AI-based cybersecurity frameworks to address these evolving threats.

Case Study: Russia's Cyber Warfare Operations

Russia's cyber capabilities have evolved significantly since the 2007 attacks on Estonia, marking a pivotal moment in the history of cyber warfare. The 2007 incident, often referred to as the first cyber war, involved a series of coordinated cyber-attacks that targeted Estonia's government, banking, and media websites, effectively crippling the country's digital infrastructure (Tikk *et al.* 2010).

Following this, Russia's cyber operations have grown in sophistication and scale. One of the most notable incidents was the 2015 Ukrainian power grid hack, where the Russian advanced persistent threat group known as 'Sandworm' used BlackEnergy 3 malware to compromise the information systems of three energy distribution companies in Ukraine. This attack resulted in power outages for approximately 230,000 consumers for 1–6 hours, marking the first publicly acknowledged successful cyber-attack on a power grid (Zetter 2016).

Another significant event was the alleged interference in the 2016 U.S. presidential election, where Russian operatives used cyber tactics to influence the election outcome. These operations included hacking into the Democratic National Committee's email servers and orchestrating disinformation campaigns on social media platforms (Mueller 2019).

AI Integration in Russian Cyber Operations

Russia has increasingly incorporated AI into its cyber arsenal, enhancing its capabilities in several key areas. AI has significantly improved Russia's ability to gather intelligence and select targets. By leveraging advanced data analytics, Russian cyber operatives can process and analyse vast amounts of data from various sources, identifying valuable information for intelligence purposes (Polyakova & Boyer 2018). This capability allows for more precise and effective targeting in cyber operations.

Machine learning algorithms play a crucial role in helping Russian cyber operations evade detection and adapt their attack patterns. These algorithms can learn from previous attempts and modify their tactics to bypass security measures. For example, AI-driven malware can continuously evolve, making it harder for traditional defence mechanisms to detect and neutralise it (Brundage *et al.* 2018).

Russia has also utilised AI in spreading disinformation and manipulating public opinion. NLP algorithms enable the creation of highly convincing fake news and social media posts. These AI-driven disinformation campaigns can influence public perception and political outcomes, as seen in the alleged interference in the 2016 U.S. presidential election (Bradshaw & Howard 2018).

The Ukraine Conflict: A Testing Ground for AI-Enhanced Cyber Warfare

The ongoing conflict in Ukraine has served as a real-world laboratory for Russia's AI-powered cyber capabilities. Russia has conducted sophisticated DDoS attacks on Ukrainian infrastructure, potentially using AI for coordination and amplification (Greenberg 2022). These attacks can overwhelm network resources, causing significant disruptions to critical services. The use of AI allows for more effective and large-scale DDoS attacks, complicating defence efforts.

AI has played a crucial role in Russia's information warfare campaigns during the Russia–Ukraine conflict. By leveraging AI, Russian operatives can create and disseminate disinformation more efficiently, manipulating public opinion and spreading confusion. These campaigns often involve the use of social media bots and deepfake technology to amplify false narratives (Jankowicz 2020). AI applications in military intelligence, surveillance, and reconnaissance have also been evident in the Russia–Ukraine conflict. AI can analyse satellite imagery and communications data to provide valuable insights for military operations. This capability enhances situational awareness and decision-making, giving Russian forces a strategic advantage (Johnson 2019). Russia's advancements in AI-enhanced cyber warfare have significant implications for global security. The integration of AI into cyber operations has increased the potential for asymmetric warfare, allowing Russia to project power beyond its conventional military capabilities. The sophisticated nature of AI-driven attacks presents challenges in attribution and deterrence, necessitating new frameworks and international cooperation to address the evolving threat landscape (Lindsay 2020).

AI Implications for Global Security

The integration of AI into cyber warfare has profound implications for global security, reshaping the dynamics of international conflict and challenging traditional notions of power and deterrence. Influence over the decision-making calculus of an adversary is now achievable across the globe and in a matter of seconds through computer and communications networks, and without the employment of 'hard power' (Sharpe *et al.* 2024; Marrone & Sabatino 2021; Harris 2017; Hunker 2010). Likewise, cyber and AI has significantly enhanced the potential for asymmetric warfare, enabling smaller powers to project substantial cyber capabilities in ways previously unimaginable (Masuhr 2019). This democratisation of cyber capabilities has the potential to disrupt traditional power balances and AI-powered cyber tools now allow nations with limited conventional military resources to compete, or at least disrupt, effectively in the digital domain. Acting as a force multiplier, AI enables small teams of cyber operators to launch large-scale, sophisticated attacks that were once the exclusive domain of major powers with extensive resources (Svenmarck *et al.* 2018). The increasing availability of AI technologies and open-source tools has lowered the barriers to entry, reducing the technical, and financial hurdles to developing advanced cyber capabilities. As a result, smaller nations and even non-state actors have become significant players in cyberspace, fundamentally altering the global digital security landscape (Karaman *et al.* 2022).

However, the sophisticated nature of AI-driven attacks introduces new challenges in attribution and deterrence. AI can be used to generate complex and evolving attack patterns, making it extremely difficult to trace the origin of cyber operations (Dilek *et al.* 2015). This obfuscation challenges traditional methods of attack attribution, complicating efforts to hold actors accountable for their actions in cyberspace. Moreover, AI-powered attacks can be designed to mimic the tactics, techniques, and procedures of other known threat actors, potentially leading to misattribution, and heightened geopolitical tensions. The rapid evolution of AI-driven threats also makes it challenging to develop effective deterrence strategies, as potential adversaries may not fully understand the capabilities they are facing, thereby increasing the risk of miscalculation and conflict escalation (Black *et al.* 2024).

Addressing the challenges posed by AI-enhanced cyber threats requires unprecedented levels of global collaboration. There is an urgent need for the international community to establish norms and guidelines for the responsible use of AI in cyber operations, including agreements on what constitutes acceptable behaviour in cyberspace. The EU has been advancing international laws and policies concerning AI and it has delivered the first comprehensive AI law designed to 'promote human-centric and trustworthy AI that prioritises the protection, health, safety, and fundamental rights' of the public (European Union 2024; European Parliament 2023). Although challenged by those who maintain a protectionist view, enhanced cooperation in the sharing of threat intelligence and best practices for AI-powered cyber defence is crucial, as it will help countries collectively improve their defensive capabilities. International partnerships in AI and cybersecurity research can accelerate the development of more robust defence mechanisms and promote a shared understanding of the evolving threat landscape (Van Den Bosch & Bronkhorst 2018). Additionally, developed nations should assist less technologically advanced countries in building their AI and cyber capabilities, ensuring a more equitable and secure global cyberspace (United Nations 2025).

Intelligence

Open-source intelligence is widely employed by military and commercial actors; this intelligence function extends to cyber (Hockenhull 2022). The traditional 'intelligence triangle' (see Figure 2.1 illustrating the hierarchies of intelligence classification) has now inverted in volume certainly, but possibly criticality too, with open-source information providing as much as 80% of reporting (OSI 2023; Williams & Blum 2018). Historically, higher classification intelligence was crucial;

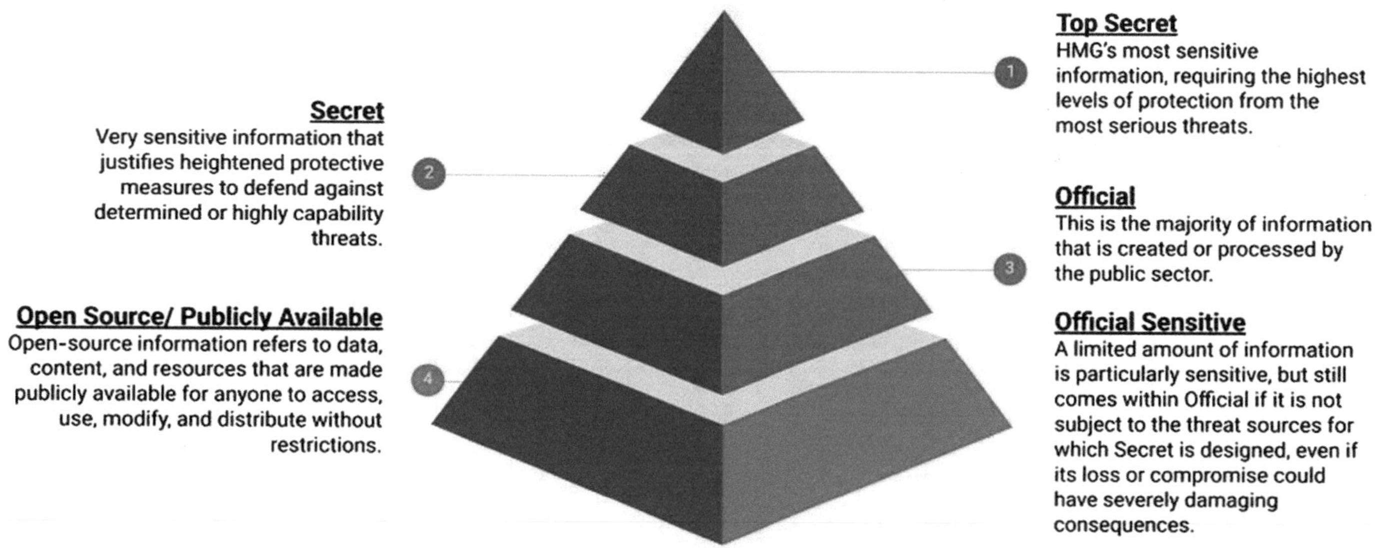

Figure 2.1 Intelligence Triangle.

Source: Ministry of Justice (MoJ). (n.d.).

however, this is arguably becoming less so as militaries and organisations leverage the vast amounts of publicly available data sets using AI.

This has empowered soldiers and civilians to create content and report enemy positions oftentimes faster and with higher fidelity than military networks. As a corollary, significant behavioural changes have been reported in enemy combatants, with Russian forces employing open-source accounts to identify vulnerabilities in both their own and in Ukrainian defensive lines (Ford 2022). Similarly, the Armed Forces of Ukraine has leveraged social media for targeting, and routinely employ civilians as sensors within the kill chain (Salerno-Garthwaite 2022). The aggregation of these various sensors, open-source or otherwise, now means that the resulting volume of data can no longer be processed manually and requires the assistance of AI capabilities.

Threats and Opportunities: Future Trends in AI-Enhanced Cyber Warfare

As AI continues to evolve rapidly, its integration into cyber warfare promises to revolutionise the landscape of digital conflicts. The future of cyber warfare is poised to be shaped by a myriad of AI-driven advancements and emerging technologies, each bringing its own set of opportunities and challenges (Singer & Friedman 2014). One of the most significant developments on the horizon is the emergence of advanced autonomous cyber agents. These AI systems are expected to operate with unprecedented levels of autonomy in cyber operations and can make complex decisions without human intervention. While this could lead to faster and more efficient cyber-attacks and defences, it also raises profound concerns about control and accountability in the digital battlefield (Libicki 2016).

AI-driven predictive analytics is another area of rapid advancement. As AI systems become more adept at analysing vast amounts of data, they will excel at predicting cyber threats before they materialise and identifying potential vulnerabilities and attack vectors with remarkable precision. This capability could significantly enhance proactive defence strategies, but it is a double-edged sword – attackers could also leverage this technology to identify and exploit weaknesses more effectively (Valeriano & Maness 2015). And while this vulnerability is not unique to AI, the 'hierarchy of needs' for supporting and enabling effective AI requires increasingly competent data analysis and intelligence capabilities across people, processes, and technologies.

The evolution of adaptive malware and defensive systems presents another fascinating frontier. AI-powered malware could evolve in real-time to evade detection, while defensive AI systems could adapt with equal agility to counter new threats. This dynamic is likely to spark an ongoing 'AI arms race' between attackers and defenders, with each side constantly striving to outmanoeuvre the other in an ever-escalating cycle of innovation (Rid 2011).

Advancements in NLP are set to reshape the landscape of social engineering attacks and defences. As NLP technologies become more sophisticated, we can expect to see highly convincing phishing attempts and disinformation campaigns.

However, these same advancements could also lead to more effective detection and prevention of such threats, thereby creating a complex interplay between offensive and defensive capabilities (Kello 2017).

The potential for AI-enabled cyber-physical attacks looms large on the horizon. AI could be leveraged to orchestrate complex attacks on critical infrastructure and other cyber-physical systems, posing significant risks to national security and public safety. This threat underscores the urgent need for advanced AI-driven protection systems to safeguard these vital assets (HSOAC 2021; Arquilla & Ronfeldt 1997). Of notable concern, the increasingly common integration of Information Technology (IT) and Operational Technology[2] (OT) presents actors with threat vectors for achieving physical outcomes through virtual access.

Emerging technologies are set to play a pivotal role in shaping the future of cyber warfare. Quantum computing, with its potential to break many current encryption methods, could fundamentally alter the cybersecurity landscape. The advent of quantum-resistant encryption will become a necessity, while the unprecedented data processing capabilities of quantum computers could enhance both offensive and defensive cyber operations (Gartzke 2013). Defensive cyber operations may be further enhanced through Edge AI. Edge AI, which involves deploying AI algorithms on local devices rather than in centralised cloud systems, promises to enhance real-time threat detection and response at the network edge. Although Edge AI could improve resilience against attacks on centralised systems, it may also pave the way for more sophisticated and distributed cyber-attacks by hostile state actors and cyber criminals (Taddeo & Floridi 2018).

The rollout of 5G and future network technologies will dramatically increase connectivity and data transfer speeds, expanding the attack surface due to the proliferation of connected devices. This will create opportunities for more sophisticated, high-bandwidth cyber-attacks, while also enhancing the capabilities of AI-driven defence systems through increased data processing capabilities (Lindsay 2020). Neuromorphic computing, with its brain-inspired chip designs, could lead to more efficient AI systems, potentially enabling more energy-efficient and faster AI-driven cyber operations. These systems could excel at processing complex unstructured data in real-time, significantly improving threat detection and response capabilities (Schmitt & Vihul 2020). Lastly, AI-enhanced biometrics could revolutionise authentication and identity verification, improving security measures against unauthorised access. However, this technology could also give rise to new types of identity-based attacks and defences, further complicating the cyber warfare landscape (Taddeo & Floridi 2018).

As we look to the future, AI-enhanced cyber warfare will be characterised by increased automation, sophistication, and the integration of an array of emerging technologies. While these advancements offer significant opportunities for improving cybersecurity, they also present new challenges, opportunities, and potential threats. Organisations and nations must stay ahead of these trends, investing heavily in research and development to maintain a competitive edge in this digital battleground. To establish a competitive edge, it is essential to recognise that first-mover advantage favours an attacker (Valeriano & Maness 2015), whilst simultaneously acknowledging the need to safeguard ethical practice in a domain that is infused with ambiguity and laden with dark corners.

The integration of quantum computing, Edge AI, and other emerging technologies will catalyse the evolution of cyber operations. This transformation will necessitate adaptation of strategies, policies, and international frameworks to address the ethical, legal, and security implications of these advanced AI-driven cyber capabilities (Singer & Friedman 2014).

Recommendations

As AI continues to reshape the landscape of cyber warfare, it is imperative for policymakers, military leaders, and cyber and information security professionals to adapt and prepare for this new battleground. The following recommendations aim to address the challenges and opportunities presented by AI in cyber warfare, ensuring a proactive, collaborative, and ethically minded approach to national security.

For Policymakers

Policymakers must take the lead in developing comprehensive AI governance frameworks. Establishing clear national and international guidelines for the ethical use of AI in cyber operations is essential (Cath *et al.* 2018). This includes working towards international agreements on the responsible development and deployment of AI in military contexts. Additionally, significant resources should be allocated to AI research and development, focusing on offensive and defensive capabilities. Encouraging public–private partnerships can leverage expertise from academia and industry, thereby promoting innovation and resilience (Svenmarck *et al.* 2018). However, this may also create opportunities for hostile state actors to exploit loopholes, steal intellectual property, and challenge security protocols. The opportunity however, to align numerous organisations around a unifying purpose, remains a government-led luxury while concurrently de-risking private sector investments ensuring maximum buy-in.

Enhancing international cooperation is another critical area. Policymakers should invest in collaborations to address global cyber threats enhanced by AI, developing mechanisms for sharing threat intelligence and best practices among allies (Maurer 2018). This must mitigate, as far as possible, the potential vulnerabilities to hostile state actors exploiting this arrangement. Addressing legal and ethical challenges is equally important. Existing laws and regulations must be updated to account for AI's role in cyber warfare, and clear accountability frameworks for AI-driven cyber operations should be established (Crootof 2019). While this is recognised as a burgeoning concern, the pace remains too slow. Technology development is occurring at an exponential rate, and is further aided by AI. At present, legislation, regulation, and policy remain far behind a position of relevance and efficacy.

For Military Leaders

Military leaders must integrate AI into military doctrine, developing new strategies and tactics that incorporate AI-enhanced cyber capabilities (Layton 2018). Training personnel in the use and countering of AI-driven cyber threats is crucial for maintaining a competitive edge. Likewise, enhancing cyber resilience is another priority. Military leaders should implement AI-driven defensive systems to protect critical military infrastructure, whilst simultaneously conducting regular AI-powered vulnerability assessments and penetration testing to fortify defences (Work & Brimley 2014).

Developing AI-enhanced early warning systems is essential for anticipating and preventing cyber-attacks. Utilising AI for predictive analysis and implementing AI-driven threat intelligence platforms for real-time situational awareness can provide a demonstrable strategic advantage (Hoadley & Lucas 2018). Given the significant geopolitical implications of the AI-cyber integration, it is advisable for military leaders to establish AI ethics committees. These committees should be tasked with overseeing the ethical implications of AI use in military cyber operations, as well as developing guidelines for maintaining human control over AI-driven decision-making processes (Horowitz *et al.* 2018).

For Cybersecurity Professionals

Cybersecurity professionals must continuously upskill in AI and machine learning to stay abreast of the latest technologies and their applications in cybersecurity. Developing expertise in AI-driven threat detection, analysis, and response is vital (Dilek *et al.* 2015). Implementing AI-enhanced security solutions, such as advanced AI-powered security information and event management (SIEM) systems, can enhance threat detection and response capabilities.

Focusing on AI model security is another critical area. Developing strategies to protect AI models from adversarial attacks and data poisoning and implementing robust testing and validation processes for the AI systems used in cybersecurity can safeguard these technologies (Papernot *et al.* 2016). Finally, it is beneficial to consider enhanced human-AI collaboration. Designing workflows that optimise the constructive collaboration between human expertise and AI capabilities, and developing interfaces that allow for effective human oversight of AI-driven security operations, can improve overall effectiveness (Shneiderman 2020).

Strategies for Developing New Technologies and Cooperative Frameworks

Establishing AI cybersecurity research centres will create dedicated facilities for the development and testing of AI-driven cyber defence technologies. Encouraging collaboration between government agencies, academic institutions, and private sector companies can drive innovation (Etzioni & Etzioni 2017) and will prove advantageous if delivered in a manner that remains attuned to threats. Moreover, developing AI testing and evaluation frameworks, with standardised methodologies for assessing the effectiveness and reliability of AI systems in cyber warfare scenarios and establishing certification processes for AI-enhanced cybersecurity tools, can ensure quality and reliability.

Creating and sustaining international AI cybersecurity alliances through appropriately validated multinational working groups focused on AI and cyber warfare, whilst simultaneously developing shared platforms for AI-driven threat intelligence sharing among allies, can enhance collective security; a Geneva Convention of the internet would support commonality of approach (Nye 2018). These benefits can be further enhanced by implementing AI governance structures with clear chains of command and decision-making processes for AI-enhanced cyber operations. Contemporaneously, the development of protocols for rapid international coordination in response to AI-driven cyber threats can streamline responses. Finally, it must be noted that promoting responsible AI development through the implementation of ethical AI principles and the development of explainable AI systems is essential for maintaining standards in the cybersecurity industry. These constructive outputs can be magnified through enhanced transparency and the resultant augmentation of trust (Dignum 2019).

Conclusion

The convergence of AI and cyber warfare will both accelerate the evolution of cyber warfare and overhaul the era of digital conflict, fundamentally reshaping the landscape of global security further towards non-kinetic activity. This chapter has explored the implications of AI in cyber warfare illustrating critical insights into its advantages and risks. Fundamentally, AI will help combatants achieve flexibility and an enhanced pace by which they can learn, adapt, and accommodate innovation; this is particularly salient within high-tech domains such as cyber.

AI enhances the speed, scale, and sophistication of cyber-attacks, significantly improving the ability to detect vulnerabilities and facilitating advanced data analysis for intelligence gathering (Brundage *et al.* 2018). Moreover, the automation and eventual autonomy of decision-making processes (supported by AI agents) in cyber operations allows for rapid and effective responses (Horowitz *et al.* 2018). However, these advancements are not without their dangers. The potential for rapid escalation of conflicts, the unpredictability of AI systems, challenges in attributing attacks to specific actors, and ethical concerns surrounding the use of autonomous cyber weapons present significant risks that must be carefully managed (Crootof 2019).

In recognition of the emerging challenges in securing cyber infrastructure, it is imperative to acknowledge that AI-powered attacks can adapt and evolve, often outpacing traditional defence mechanisms (Dilek *et al.* 2015). The complexity of securing and auditing AI systems introduces new vulnerabilities, while the enhanced capabilities of AI also amplify the threat of sophisticated social engineering attacks. It is advisable for both policymakers and practitioners alike to assimilate the proposition that AI minimises human errors but also provides a key tool that attackers can leverage for social engineering attacks such as phishing.

The resultant dichotomy necessitates continuous updates and training of defence systems. Crucially, it also requires personnel to adopt a proactive stance by keeping pace with evolving threats and preventing malicious intrusion on critical networks (Svenmarck *et al.* 2018).

A case study of Russia's cyber warfare operations illustrated many of these dynamics in action. Russia has demonstrated sophisticated AI-enhanced cyber capabilities, particularly in its use of disinformation campaigns, advanced data analytics, and adaptive attack patterns (Maurer 2018). The ongoing conflict in Ukraine has served as a critical testing ground for these capabilities, highlighting the potential for AI-driven cyber operations to disrupt critical infrastructure and manipulate public opinion (Kello 2017).

Cyberspace is the new frontline for national defence and security with significant disparities in approaches and capabilities among major powers. While the United Kingdom has established a strong foundation in cybersecurity, it faces ongoing challenges in maintaining parity with the rapid advancements made by other nations. The strategic importance of cyber capabilities is universally recognised, yet the resources allocated and methodologies employed vary considerably between countries. As the digital realm continues to evolve, the UK and others must remain vigilant and adaptive in its cyber strategies (such as security, AI, and doctrinal approaches to cyber warfighting) to effectively protect its national interests and maintain its position in the international cyber arena. This underscores the need for continued investment, innovation, and international cooperation to address cyber threats and opportunities and effectively modulate competition and conflict in cyberspace.

The integration of AI into cyber warfare also raises significant legal and ethical considerations. The lack of clear international norms and definitions complicates the establishment of accountability frameworks for AI-driven cyber operations. Ethical dilemmas surrounding autonomy, moral responsibility, bias, and transparency further complicate the deployment of AI in military contexts (Dignum 2019). These challenges necessitate a concerted effort to ensure compliance with international humanitarian law while safeguarding civil liberties and privacy rights and balancing innovation and regulation of AI capabilities.

Potential developments in AI and emerging technologies, such as quantum computing and Edge AI, will further transform the landscape of cyber warfare (Wallden & Kashefi 2019). Advanced autonomous cyber agents, predictive analytics, and adaptive malware are prominent examples of innovations that will shape the future of this field (Apruzzese *et al.* 2018). As these technologies evolve, they will require continuous adaptation of strategies, policies, and international frameworks to address the ethical, legal, and security implications of AI-enhanced cyber capabilities (Lindsay 2020).

The integration of AI into cyber warfare represents a change in basic assumptions in global security dynamics. While AI offers unprecedented capabilities, it exposes notable vulnerabilities and introduces significant risks that must be addressed through proactive and collaborative efforts. Policymakers, military leaders, and cybersecurity professionals must prioritise the development of comprehensive governance frameworks, invest in AI research and development, and enhance international cooperation (Etzioni & Etzioni 2017). In doing so, allied partners can collectively navigate the cyber battleground with confidence, while maintaining operational resilience and the stability and security of the digital domain (Shneiderman 2020).

Notes

1 Proprietary AI models, developed by tech giants like IBM, Google, and Microsoft, leverage vast resources to create high-performance systems with advanced algorithms and specialised hardware. These models often excel due to extensive training on large datasets. In contrast, open-source AI models, such as OpenAI and open-source GPT variants, are collaboratively developed by a global community. This approach ensures transparency, enabling thorough scrutiny of the models' inner workings and supports 'explainability'. Such openness is particularly valuable in fields where regulatory compliance and interpretability are critical. Both proprietary and open-source models have their strengths, catering to different needs in the AI landscape. Proprietary models often lead in performance, while open-source alternatives provide transparency and community-driven innovation.

2 Operational technology refers to the hardware and software systems used to monitor, control, and manage physical devices, processes, and infrastructure in industrial and critical environments (Hashemi-Pour *n.d.*).

References

The Alan Turing Institute. n.d.. Data Science and AI Glossary. Available at: www.turing.ac.uk/news/data-science-and-ai-glossary

Amster, Alan. 2024. The Role of Artificial Intelligence in Combating Phishing and Other Cybercrimes. ALLSTARTSIT. Accessed August 6, 2024. Available at: www.allstarsit.com/blog/the-role-of-artificial-intelligence-in-combating-phishing-and-other-cybercrimes#:~:text=AI%2Dpowered%20tools%20can%20automate,and%20speed%20up%20threat%20detection.

Apruzzese, Giovanni, Mauro Colajanni, Luca Ferretti, Alessandro Guido, and Mirco Marchetti. 2018. On the Effectiveness of Machine and Deep Learning for Cyber Security. In 2018 10th International Conference on Cyber Conflict (CyCon), 371–390. IEEE.

Arquilla, John, and David Ronfeldt. 1997. *In Athena's Camp: Preparing for Conflict in the Information Age*. RAND Corporation.

Black, James, Mattias Eken, Jacob Parakilas, Stuart Dee, Conlan Ellis, Kiran Suman-Chauhan, Ryan Bain, Harper Fine, Maria Aquilino, Melusine Lebret, and Palicka, A. 2024. Strategic Competition in the Age of AI. Available at: www.rand.org/pubs/research_reports/RRA3295-1.html.

Bradshaw, Samantha, and Philip N. Howard. 2018. *Challenging Truth and Trust: A Global Inventory of Organized Social Media Manipulation*. Oxford Internet Institute.

Brundage, M., Avin, S., Clark, J., Toner, H., Eckersley, P., Garfinkel, B., Dafoe, A., Scharre, P., Zeitzoff, T., Filar, B., Anderson, H., Roff, H., Allen, G. C., Steinhardt, J., Flynn, C., HEacute;igeartaigh, S. Oacute;., Beard, S., Belfield, H., Farquhar, S., *et al.* (2018). *The Malicious Use of Artificial Intelligence: Forecasting, Prevention, and Mitigation.* Apollo – University of Cambridge Repository. https://doi.org/10.17863/CAM.22520.

Buchanan, Ben. 2020. *The Hacker and the State: Cyber Attacks and the New Normal of Geopolitics*. Cambridge, MA: Harvard University Press.

Cath, Corinne, Sandra Wachter, Brent Mittelstadt, Mariarosaria Taddeo, and Luciano Floridi. 2018. Artificial Intelligence and the 'Good Society': the US, EU, and UK Approach. *Science and Engineering Ethics* 24 (2): 505–528.

Chesney, Robert, and Danielle Citron. 2019. Deepfakes and the New Disinformation War: The Coming Age of Post-Truth Geopolitics. *Foreign Affairs* 98 (1): 147–155.

Crootof, Rebecca. 2019. Artificial Intelligence, Autonomous Weapons, and the Future of Warfare. *Harvard National Security Journal* 10: 135–187.

Cullen, P., and Reichborn-Kjennerud, E. (2017). MCDC Countering Hybrid Warfare Project: A Multinational Capability Development Campaign Project. Available at: https://assets.publishing.service.gov.uk/media/5a8228a540f0b62305b92caa/dar_mcdc_hybrid_warfare.pdf.

Dewar, Robert S. 2018. *Cyber Security and Cyber Defence in the European Union: Opportunities, Synergies and Challenges*. EU Institute for Security Studies.

Dignum, Virginia. 2019. *Responsible Artificial Intelligence: How to Develop and Use AI in a Responsible Way*. Cham, Switzerland: Springer Nature.

Dilek, Selma, Hüseyin Çakır, and Mustafa Aydın. 2015. Applications of Artificial Intelligence Techniques to Combating Cyber Crimes: A Review. *International Journal of Artificial Intelligence & Applications* 6 (1): 21–39.

Etzioni, Amitai, and Oren Etzioni. 2017. Incorporating Ethics into Artificial Intelligence. *The Journal of Ethics* 21 (4): 403–418.

European Parliament. 2023. EU AI Act: First Regulation on Artificial Intelligence. Accessed September 14, 2024. Available at: www.europarl.europa.eu/topics/en/article/20230601STO93804/eu-ai-act-first-regulation-on-artificial-intelligence.

European Union. 2024. Regulation (Eu) 2024/1689 of the European Parliament and of the Council. Available at: https://eur-lex.europa.eu/legal-content/EN/TXT/?uri=CELEX%3A32024R1689.

Ferrara, Emilio, Onur Varol, Clayton Davis, Filippo Menczer, and Alessandro Flammini. 2016. The Rise of Social Bots. *Communications of the ACM* 59 (7): 96–104.

Fitzgerald, Anna, and Emily Bonnie. 2024. AI in Cybersecurity: How It's Used + 8 Latest Developments. Secureframe. Accessed August 6, 2024. Available at: https://secureframe.com/blog/ai-in-cybersecurity.

Ford, Matt. 2022. The Smartphone as Weapon part 1: The New Ecology of War in Ukraine. Available at: www.academia.edu/75845985/The_Smartphone_as_Weapon_part_1_the_new_ecology_of_war_in_Ukraine

Fortinet *n.d.* What Is Cyberwarfare? Available at: www.fortinet.com/de/resources/cyberglossary/cyber-warfare.

Gartzke, Erik. 2013. The Myth of Cyberwar: Bringing War in Cyberspace Back Down to Earth. *International Security* 38 (2): 41–73.

Geist, Edward, and Andrew J. Lohn. 2018. How Might Artificial Intelligence Affect the Risk of Nuclear War? Santa Monica, CA: RAND Corporation.

Goodfellow, Ian, Jonathon Schlens, and Christian Szegedy. 2015. Explaining And Harnessing Adversarial Examples. Available at: https://arxiv.org/pdf/1412.6572.

Greenberg, Andy. 2022. Russia's Sandworm Hackers Have Built a Botnet of Firewalls. *Wired*, June 16, 2022.

Hallaq, Bilal, Tiia Somer, Anna-Maria Osula, Kim Ngo, and Timothy Mitchener-Nissen. 2017. Artificial Intelligence within the Military Domain and Cyber Warfare. Proceedings of the 16th European Conference on Cyber Warfare and Security (ECCWS 2017), Dublin, Ireland, June 29–30. Academic Conferences and Publishing International Limited. Accessed August 6, 2024. https://wrap.warwick.ac.uk/94297/.

Harris, S. 2017. China Reveals Its Cyberwar Secrets. The Daily Beast. Available at: www.thedailybeast.com/china-reveals-its-cyberwar-secrets.

Hasemi-Pour, C. 2024. Operational Technology. Available at: www.techtarget.com/whatis/definition/operational-technology.

Heitzenrater, Chad. 2023. Cyber Attacks Reveal Uncomfortable Truths About U.S. Defenses. RAND Corporation. Accessed August 6, 2024. Available at: www.rand.org/pubs/commentary/2023/09/cyber-attacks-reveal-uncomfortable-truths-about-us.html.

Hoadley, Daniel S., and Nathan J. Lucas. 2018. *Artificial Intelligence and National Security*. Congressional Research Service.

Hockenhull, J. 2022. *How Open-Source Intelligence Has Shaped the Russia-Ukraine War*. Available at: www.gov.uk.

Homeland Security Operational Analysis Center (HSOAC). 2021. Cybersecurity. Available at: www.rand.org/pubs/corporate_pubs/CPA1329-2.html

Horowitz, Michael C., Gregory C. Allen, Elsa B. Kania, and Paul Scharre. 2018. *Artificial Intelligence and International Security*. Washington, D.C.: Center for a New American Security.

Hunker, J. (2010). Cyber War and Cyber Power: Issues for NATO Doctrine. NATO Defense College. Available at: www.jstor.com/stable/resrep10354.

IBM. n.d. Artificial Intelligence (AI) Cybersecurity. Accessed August 6, 2024. Available at: www.ibm.com/ai-cybersecurity.

Imperva. n.d. What Is Cyber Warfare? Types, Examples & Mitigation. Accessed August 6, 2024. Available at: www.imperva.com/learn/application-security/cyber-warfare/.

Jankowicz, Nina. 2020. *How to Lose the Information War: Russia, Fake News, and the Future of Conflict*. London: I.B. Tauris.

Johnson, James. 2019. Artificial Intelligence & Future Warfare: Implications for International Security. *Defense & Security Analysis* 35 (2): 147–169.

Karaman, M., H. Çatalkaya, and C. Aybar. 2022. Institutional Cybersecurity from Military Perspective. In *Artificial Intelligence and Machine Learning Applications in Civil, Mechanical, and Industrial Engineering*, 237–254. IGI Global.

Kello, Lucas. 2017. *The Virtual Weapon and International Order*. Yale University Press.

Kott, Alexander, David S. Alberts, and Cliff Wang. 2015. Will Cybersecurity Dictate the Outcome of Future Wars? *Computer* 48 (12): 98–101.

Layton, Peter. 2018. Algorithmic Warfare: Applying Artificial Intelligence to Warfighting. *Air & Space Power Journal* 32 (1): 47–60.

Lewis, James Andrew. 2022. *Cyber War and Ukraine*. Center for Strategic and International Studies. Available at: www.csis.org/analysis/cyber-war-and-ukraine.

Libicki, Martin C. 2016. *Cyberspace in Peace and War*. Annapolis, MD: Naval Institute Press.

Lindsay, Jon R. 2020. Artificial Intelligence and International Security: The Long View. *Ethics & International Affairs* 34 (2): 139–159.

Lindsay, Jon R., Tai Ming Cheung, and Derek S. Reveron. 2015. *China and Cybersecurity: Espionage, Strategy, and Politics in the Digital Domain*. New York: Oxford University Press.

Marrone, A., and Sabatino, E. (2021). Cyber Defence in NATO Countries: Comparing Models. Istituto Affari Internazionali (IAI). Available at: www.jstor.org/stable/resrep28807.

Masuhr, N. 2019. AI in Military Enabling Applications. *CSS Analyses in Security Policy* 251: 1–4.

Maurer, Tim. 2018. *Cyber Mercenaries: The State, Hackers, and Power*. Cambridge New York: University Press.

Ministry of Defence (MoD). 2022. *Defence Artificial Intelligence Strategy*. Available at: Defence_Artificial_Intelligence_Strategy.pdf (publishing.service.gov.uk).

Ministry of Justice (MoJ). (n.d.) Government Classification Scheme. Available at: https://security-guidance.service.justice.gov.uk/government-classification-scheme/#government-classification-scheme

Mueller, Robert S. 2019. *Report On the Investigation into Russian Interference in the 2016 Presidential Election*. Washington, D.C.: U.S. Department of Justice.

Nye, Joseph S., Jr. 2018. How Will New Cybersecurity Norms Develop? Available at: www.aspistrategist.org.au/how-will-cybersecurity-norms-develop/.

Open-Source Initiative (OSI). 2023. The 2023 State of Open-Source Report Confirms Security as Top Issue. Available at: The 2023 State of Open Source Report confirms security as top issue – Open Source Initiative.

Palo Alto Networks. 2024a. What Is Security Automation? Accessed August 6, 2024. Available at: www.paloaltonetworks.com/cyberpedia/what-is-security-automation.

Palo Alto Networks. 2024b. AI Risk Management Framework. Accessed 14 August, 2024. Available at: www.paloaltonetworks.co.uk/cyberpedia/ai-risk-management-framework.

Papernot, Nicolas, Patrick McDaniel, Arunesh Sinha, and Michael Wellman. 2016. Towards the Science of Security and Privacy in Machine Learning. arXiv preprint arXiv:1611.03814.

Petersen, N. 2023. The Chinese Communist Party's Theory of Hybrid Warfare. Available at: www.understandingwar.org/backgrounder/chinese-communist-partys-theory-hybrid-warfare.

Polyakova, Alina, and Spencer P. Boyer. 2018. *The Future of Political Warfare: Russia, the West, and the Coming Age of Global Digital Competition*. Brookings Institution. www.brookings.edu/wp-content/uploads/2018/03/fp_20180316_future_political_warfare.pdf.

Rid, Thomas. 2011. Cyber War Will Not Take Place. *Journal of Strategic Studies*, 35 (1): 5–32. https://doi.org/10.1080/01402390.2011.608939.

Rudin, Cynthia. 2019. Stop Explaining Black Box Machine Learning Models for High Stakes Decisions and Use Interpretable Models Instead. *Nature Machine Intelligence* 1 (5): 206–215. www.nature.com/articles/s42256-019-0048-x.

Salerno-Garthwaite, Andrew. 2022. OSINT in Ukraine: Civilians in the Kill Chain and Information Space. *Global Defence Technology* 137. Available at: nridigital.com.

Scharre, Paul. 2018. *Army of None: Autonomous Weapons and the Future of War*. New York: W. W. Norton & Company.

Schmitt, Michael N., and Liis Vihul. 2020. International Law and Cyber Attacks: Sony v. North Korea. *Journal of National Security Law & Policy* 9 (1): 213–242.

Sharpe, Jack, Markos Trichas, and Damian Terrill. 2024. Culture: A Sixth Domain and the Introduction of the 'C6ISRT' Framework. *Defence Studies* 25 (1): 22–46. https://doi.org/10.1080/14702436.2024.2397520.

Shneiderman, Ben. 2020. Human-Centered Artificial Intelligence: Reliable, Safe & Trustworthy. *International Journal of Human–Computer Interaction* 36 (6): 495–504.

Singer, P. W., and Allan Friedman. 2014. *Cybersecurity and Cyberwar: What Everyone Needs to Know*. Oxford University Press.

Sontan, A., and S. Samuel. 2024. The intersection of Artificial Intelligence and cybersecurity: Challenges and Opportunities. *World Journal of Advanced Research and*

Reviews. Accessed September 14, 2024. Available at: https://wjarr.com/sites/default/files/WJARR-2024-0607.pdf.

Steinhardt, Jacob, Pan Wei Koh, and Percy Liang. 2017. Certified Defenses for Data Poisoning Attacks. Available at: https://arxiv.org/pdf/1706.03691.

Svenmarck, P., L. Luotsinen, M. Nilsson, and J. Schubert. 2018. Possibilities and Challenges for Artificial Intelligence in Military Applications. In Proceedings of the NATO Big Data and Artificial Intelligence for Military Decision Making Specialists' Meeting, 1–16.

Szabadföldi, I. 2021. Artificial Intelligence in Military Application – Opportunities and Challenges. *Land Forces Academy Review* 26 (2): 157–165.

Taddeo, Mariarosaria, Luciano Glorioso, and Luciano Floridi. 2018. Regulate Artificial Intelligence to Avert Cyber Arms Race. Available at: www.nature.com/articles/d41586-018-04602-6.

Tikk, Eneken, Kadri Kaska, and Liis Vihul. 2010. *International Cyber Incidents: Legal Considerations*. NATO Cooperative Cyber Defence Centre of Excellence.

Tramèr, Florian, Fan Zhang, Ari Juels, Michael K. Reiter, and Thomas Ristenpart. 2016. Stealing Machine Learning Models via Prediction APIs. In 25th USENIX Security Symposium (USENIX Security 16), 601–618.

United Nations. 2025. Global Collaboration for Inclusive and Equitable AI. Available at: https://unctad.org/system/files/official-document/tir2025ch5_en.pdf.

Valeriano, Brandon, and Ryan C. Maness. 2015. *Cyber War versus Cyber Realities: Cyber Conflict in the International System*. Oxford University Press: New York.

Van Den Bosch, K., and A. Bronkhorst. 2018. Human-AI Cooperation to Benefit Military Decision Making. In *Proceedings of the NATO STO-MP-SCI-300 Symposium on Artificial Intelligence for Military Multi Domain Operations in 2040*, 1–14.

Wallden, Petros, and Elham Kashefi. 2019. Cyber Security in the Quantum Era. *Communications of the ACM* 62 (4): 120–129.

Williams, Heather, and Ilana Blum. 2018. Defining Second Generation Open-Source Intelligence (OSINT) for the Defense Enterprise. Available at: www.rand.org/pubs/research_reports/RR1964.html .

Work, Robert O., and Shawn Brimley. 2014. *20YY: Preparing for War in the Robotic Age*. Center for a New American Security.

Zetter, Kim. 2016. Inside the Cunning, Unprecedented Hack of Ukraine's Power Grid. Accessed August 6, 2024. Available at: www.wired.com/2016/03/inside-cunning-unprecedented-hack-ukraines-power-grid.

3 Fighting for Influence

The Promise of Artificial Intelligence

Andrew N. Liaropoulos

Introduction

Influence operations (IOs) are as old as military history. The use of information to deceive or manipulate adversaries and civilian populations has always been an important element of warfare. Throughout history, technology has been a factor that significantly transformed these operations (Taylor 2003; Fridman et al. 2022). Over the past two decades, the advent of information and communication technologies (ICTs) and the proliferation of social media platforms have facilitated the rapid dissemination of messages to a broader audience, thus enhancing the speed and effectiveness of IOs. The weaponization of social media is targeting our attention. Posting photos and videos, sharing tweets and likes, providing convincing stories, and spreading fake news are the new weapons in the information-intensive battle for our hearts and minds (Liaropoulos 2023).

Artificial intelligence (AI) has triggered a further evolution in the practice of such operations. AI-powered tools create vast amounts of content quickly. This includes text, images, and videos that are used to spread narratives, influence opinions, and even create fake personas. Chatbots and conversational AI can engage with individuals online, spread information, and shape public discussions. The ability to create artificial individuals, fabricate realistic videos, and manufacture a false consensus on key issues, suggests that the battle for our hearts and minds may be entering a new phase.

Such capabilities are of paramount importance for militaries that conduct IOs. After all, the creation of content that is tailored to specific audiences and demographics, is one of the most critical components of an influence campaign. Yet, AI-powered tools offer more than content generation, such as the fabrication of false identities. One of the primary challenges in IOs is the construction of credible avatars to disseminate information. A key concern in this context is the creation of identities that are both credible and do not compromise the security of the personnel conducting these operations. Moreover, the increasing accessibility of AI-powered technologies to lower-ranked military powers and nonstate actors, like ISIS and Hezbollah, will create further asymmetries in the future battlefield (Mazzucchi 2022, 6–7).

DOI: 10.4324/9781003520160-3

The purpose of this chapter is to explore the potential that AI offers for implementing IOs. The reader should note that this is up to a point a speculative task since both AI and IOs are constantly evolving. Thus, the aim is to not only identify the main areas where AI-powered tools impact the execution of such operations, accompanied by real-world examples, but also identify potential limitations. The layout of this chapter is the following. First, we discuss the concept of IOs. There is an ongoing debate on how to define such operations, since they are associated with concepts like information operations, psychological operations, disinformation, propaganda, and of lately, cognitive warfare. Second, having sorted out the conceptual framework, we analyze how AI-powered tools alter the running of IOs. We apply the actors, behaviors, and content (ABC) framework to describe how AI transforms the conduct of IOs. This involves among others, the use of language models to generate content, deepfake technology to manipulate public opinion, and automated bots for amplifying messages. Third, we explore future trends and limitations regarding the use of AI-powered tools in influence campaigns. This chapter ends with a summary of our research findings.

Influence Operations 2.0

Information is an essential element of power used throughout history to exert influence on other actors and achieve political objectives. The current information environment is fragmented and dominated by powerful information platforms. It is shaped by factors like the ability to collect a large amount of data, the rise of a trolling culture, the virality on social media, and the impact of self-reinforcing echo chambers (Mazarr et al. 2019, 19). A review of the literature demonstrates that there are various terms used to conceptualize how the current information environment is shaped to influence the opinions of a targeted audience. Terms like propaganda, strategic communication, misinformation, disinformation, public diplomacy, information warfare, psychological warfare, perception management, foreign information manipulation and interference, and cyber-enabled information operations are widely used in the public domain. Adding to that, IOs are closely associated with cognitive warfare and political warfare, which are all perceived as elements of hybrid warfare (Wanless and Pamment 2019).

Our attempt here is not to engage in a definitional debate and add conceptual confusion, but rather to clarify why the term IOs is the preferred one concerning the aims of our research. First, our focus is on how an actor can utilize AI-powered tools to exert influence on a targeted audience; thus, whether the content spread is true or false (e.g. propaganda, misinformation, disinformation, fake news) is irrelevant. Likewise, our emphasis is on detecting the means and methods of exerting influence and not on detecting the identity of the actor applying these means and methods (e.g. propaganda, public diplomacy). In other words, does truth matter, in terms of describing the content used in the context of an influence campaign (Wanless and Pamment 2019, 3–4)?

Second, although IOs encompass the use of information assets, the terms information warfare and information operations can be confusing since they include, among others, electronic warfare, cyber operations, psychological operations, military deception, and operations security and are perceived by some as actions restricted in wartime. Information operations aim to influence, disrupt, and corrupt adversarial human and automated decision-making while protecting our own. IOs refer to actions taken to affect behaviors, decisions, or perceptions of targeted audiences to achieve certain objectives. The former operations are broader and involve a wide range of military and technical means, like cyber-attacks, whereas the latter are designed to shape perceptions and decisions via informational means (Whyte et al. 2021) and operate during both wartime and peacetime. Semantics are important. Elements of information operations such as psychological operations and electronic warfare are distinctly military terms, even though they function very similarly to activities and efforts also undertaken by nonmilitary agencies. Creating a term to cover all these disparate activities is difficult and perhaps pointless (Armistead 2010, 94–95).

Third, using the cyber prefix and coining terms like cyber-enabled information operations or cyber-IOs does not offer any added value, apart from stating the utility of cyberspace as an enabler of such operations (Lin and Kerr 2021). Cyberspace typically consists of three layers, the physical, the logical, and the social layer. Even though IOs target the human domain in the social layer, they rely on the other two as well (e.g. the conduct of cyber-attacks). A cyber-attack targets the information infrastructure whereas IOs utilize – weaponize – information to affect the perception of the target and thereby its decision-making mechanism (Ramluckan and van Niekerk 2019, 68).

Thus, for the present analysis, we consider IOs as synchronized actions, which utilize a range of resources – diplomatic, economic, informational, intelligence, psychological, communication, military, technical, and cultural – to shape the attitudes, behavior, and decisions of a specific target audience, to align them with the objectives of the influencer (Gregor and Mlejnková 2021, 21–22). IOs come in many shapes and forms. They involve persuasion, coercion, and manipulation. Persuasion involves appealing openly to reason and conscious deliberation, aiming for the targeted audience to understand and independently decide to change their behavior. In contrast, coercive IOs restrict the choices of the audience, leaving them with little or no meaningful options, thereby forcing compliance rather than stimulating voluntary change. Manipulation, however, goes further by seeking to undermine or seize control of the audience's decision-making capacity. It operates through psychological tactics, exploiting biases and heuristics to achieve its ends subtly and often without the audience's awareness (Cristiano and van den Berg 2023, 99–100). The *modus operandi* may include the dissemination of fake content and the use of information to encourage specific behavior, or the paralysis of the adversary's decision-making processes by intentionally promoting discursive polarization. Below are recent examples of AI-enabled IOs.

Over the past years, Facebook, X (formerly known as Twitter), WhatsApp, Telegram, and other information platforms have identified and removed hundreds of IOs, operating from different countries. These operations include among others the use of deepfakes, automated web data scraping, and inauthentic accounts or bot networks that engage in sophisticated social engineering campaigns (Juršėnas et al. 2021; Virtual Manipulation Brief 2024). The sociopolitical context of such operations ranges, from domestic policy issues like the rights of Black Americans (e.g. the Black Live Matters movement) and election interference (e.g. the 2016 US Presidential Elections, the 2017 French Presidential Elections), to the support of military operations (e.g. the 2022 ongoing Russia–Ukraine conflict).

On March 16, 2022, the Ukrainian television channel Ukraine 24 was reportedly hacked by pro-Russian actors, resulting in the broadcast of a written message falsely attributed to President Zelensky, urging Ukrainian soldiers to surrender. On the same day, deepfake videos featuring President Zelensky were circulated on the messaging platform Telegram, and the Russian social media platform Vkontakte. The deepfake video was relatively unsophisticated, featuring poor voice samples and technical flaws in the animation, leading to its swift debunking and thus having minimal impact on the Ukrainian public. However, the use of deepfake videos involving high-profile political leaders during wartime signifies a new way of exerting influence (Mazzucchi 2022, 14). A similar case is the fake BBC video claiming that a Ukrainian politician sold arms to Hamas. This video also falsely claimed that an online investigative group called Bellingcat was supporting these allegations. Although this video was debunked, it consumed significant resources and diverted attention from countering other disinformation products aimed at Ukraine (Reuters Fact Check 2023; Karalis 2024, 521).

In May 2024, OpenAI identified and disrupted five IOs involving actors from Russia, China, Iran, and Israel. These operations were using its AI models – including ChatGPT, to generate short comments and longer articles in a range of languages, making up names and bios for social media accounts, conducting open-source research, debugging simple code, and translating and proofreading texts, to deceptively influence public opinion and political discourse (OpenAI 2024). Two operations originated from Russia. The first one, nicknamed by Open AI as Bad Grammar focused on generating, via GhatGPT, comments about the Russia–Ukraine war as well as debug code for running a Telegram bot and creating short, political comments in Russian and English that were then posted on the platform. Bad Grammar targeted mainly Ukraine, Moldova, the Baltic States, and the United States (Farooq 2024; French 2024). The second one, labeled Doppelganger, used AI to generate comments in English, French, German, Italian, and Polish that were then posted on X, translate and edit articles in English and French that were posted on websites linked to this operation, and convert news articles into Facebook posts. The inauthentic news articles spread narratives that undermined Ukraine's strength and stability and questioned its relations with its Western allies (Recorded Future 2023).

Spamouflage is the name given to a Chinese operation that used AI to research public social media activity and generate texts, mainly in Chinese, English, Japanese, and Korean that were then posted across platforms including X, Medium, and Blogspot (OpenAI 2024). Likewise, the International Union of Virtual Media used GhatGPT mostly for proofreading, headline, and tag generation. The content generated was pro-Iran and against the US and Israel. The last case is that of Zero Zeno, an influence campaign run by the Israeli political campaign management firm called STOIC. This operation involved AI-generated social media posts across Instagram, Facebook, and X, attempting to sway opinion on various topics including the Israel–Hamas war and US involvement in Middle East conflicts, and targeted audiences in the US and Israel (Farooq 2024; French 2024). According to an OpenAI report, "These trends reveal a threat landscape marked by evolution, not revolution. Threat actors are using our platform to improve their content and work more efficiently. But so far, they are still struggling to reach and engage authentic audiences" (OpenAI 2024, 7). The above examples only slightly reveal the true potential of AI. Thus, the following section illustrates in detail how AI affects the art and science of influence.

The Use of AI-Powered Capabilities in IOs

The potential impact of AI-powered tools on the conduct of IOs is approached by applying the ABC framework. "A" stands for the actor(s) waging the influence campaign, "B" for the deceptive behaviors, and "C" for the content itself. We have to consider all three dimensions, because one of them may remain authentic within a broader manipulative campaign (François 2019). For instance, authentic content may be artificially amplified through paid or automated engagement, or by actors whose identities are deceptive. Likewise, authentic actors may employ inauthentic automated techniques. Therefore, when considering the potential impact of AI's future influence activities, it is essential to examine its capacity to transform each of these dimensions (Goldstein et al. 2023, 22).

A Plethora of Actors

Concerning actors, a key constraint for conducting influence campaigns is the associated cost. Although social media has lowered the expense of reaching broader and more diverse audiences, most disinformation efforts still require the creation of numerous fake personas, advanced automation, and a consistent flow of relevant content. The advent of AI further reduces these costs by automating content production, streamlining the creation of fabricated personas, and generating culturally tailored outputs that are less likely to exhibit detectable signs of inauthenticity (Goldstein 2023, 23). Apart from reducing cost, AI-powered tools lead to a wider pool of potential actors waging IOs. State and nonstate actors outsource their influence campaigns to private firms that offer influence as a service. The industry offers a vast array of AI-powered capabilities to affect an audience's emotions, ideas, and behaviors, to advance a state or nonstate actor's objectives (Briant and Bakir 2024). Even if AI companies place restrictions on who can access their models, public relations and digital marketing firms will most probably be granted access to such capabilities since they are considered legitimate entities. An obvious side effect of the above is that by outsourcing some elements of an operation to a third party, influencers receive plausible deniability in return (Sedova 2021, 14–15). In the case of the Russian IOs in Syria during the period 2014–2018, orchestrated by the military intelligence agency GRU and the private company Internet Research Agency (IRA), the use of digital mercenaries made attribution a significant challenge, thus offering the Kremlin plausible deniability for its actions (DiResta et al. 2021).

Deceptive Behaviors

The key behavioral shift resulting from using large language models (LLMs) in IOs, lies in the replacement or augmentation of human writers in content generation. By substituting human authors with LLMs or employing them in human–machine teams, the cost of conducting influence campaigns is significantly reduced, while the scalability of operations is dramatically enhanced. This includes campaigns such as mass messaging on social media or the production of long-form news articles for unattributable websites. In terms of operational security, if a campaign is executed with a smaller staff, this translates into fewer potential leaks and moles (Goldstein and Sastry 2023, 3).

Beyond merely generating text, AI-powered models can enhance existing tactics, techniques, and procedures used in IOs, such as cross-platform testing. The latter involves the process where content is initially tested on one platform to estimate audience response before being disseminated to others. This process can be optimized by generative models. Moreover, as LLMs continue to advance, agents of influence may soon be able to utilize demographic data to produce more targeted and persuasive content. Unlike human writers, who are limited by time and capacity, LLMs could generate unique articles tailored to specific combinations of demographics, which would be impractical for human authors to accomplish at a large scale. The effectiveness of this approach will depend on the persuasiveness of AI-generated text and the degree to which highly personalized content outperforms traditional, less-targeted human-written articles (Goldstein 2023, 24–25). Finally, another key example of the potential of AI is the capacity of LLMs for sustained dialogue. Chat-based systems could engage in prolonged, personalized conversations with individual targets, providing the opportunity for influencers to deploy customized chatbots to persuade users on a one-on-one basis (Goldstein and Sastry 2023, 4–5).

Fake and Automated Content

When evaluating the impact of AI-powered tools on the content of an influence campaign, it is essential to consider how the content is generated and distributed to the targeted audience. Automatic web data scraping serves multiple purposes, enabling the collection of vast amounts of human-generated data in various formats, including text, images, video, and audio. This data can be exploited by AI-powered tools to replicate organic behavior, such as deploying media bots that post fragments of scraped content from the web. Influence actors may use this data to train generative models that emulate real-world content across different mediums. Additionally, influence actors exploit machine translation services to automatically translate scraped content, such as material from websites promoting divisive narratives, into the language of their target audience. Moreover, AI algorithms have the potential to enhance existing web crawlers by making them more resilient to changes in website code and capable of mimicking human-like activity patterns to evade detection by automation defenses. Web scraping remains a cost-effective method for acquiring large volumes of human-generated content, which can then be used to mask the operations of bot networks, disseminate divisive information, and generate extensive datasets for training malicious text-generating models (Juršėnas et al. 2021, 9–10).

Indicative of the above is the research conducted in 2021 at Georgetown University's Center for Security and Emerging Technology. The researchers tested for six months GPT-3's persuasiveness and ability to generate tailored messages around a false narrative. The chosen topics were the withdrawal of troops from Afghanistan and sanctions on China. The researchers presented volunteers with sample tweets generated by GPT-3 on these two topics and discovered that the AI-generated messages had a significant influence on the participants' opinions (Buchanan et al. 2021).

Recent advances in LLMs – like the OpenAI's GPT series, the Meta's LLaMA models, and Google's Gemini – that mimic human language are important milestones for the conduct of IOs. These developments thrived, due to the development of larger and more versatile model architectures, the expansion of dataset sizes, and the substantial investments by technology companies in enhancing computational resources for model training (IISS 2023). Content – text generated by LLMs, such as GPT-4 – can significantly influence readers' beliefs, including on sensitive political issues, with effects comparable to those of human-authored influence tools. As these models continue to advance there is the possibility that LLMs will soon be capable of accurately replicating marginal ideologies and avoiding common errors made by human operators such as mistranslating idiomatic expressions or using methods like copying and pasting identical content across multiple accounts to save time (Goldstein 2023, 26–28). The ability to provide linguistic and cultural context of the targeted audience and thereby produce linguistically distinct messaging, offers strong incentives for influence actors to integrate LLMs into their operations. That was the case with CopyCop, a Russian influence campaign identified in early 2024, which used LLMs to generate content that promoted pro-Russian narratives concerning the war in Ukraine and the Israel–Hamas conflict. The campaign modified content from mainstream media outlets like BBC, Fox News and Al-Jazeera, to shape public perception in the US, UK, and France. The operation tailored narratives to specific audiences and produced more than 19,000 articles per month. These numbers demonstrate the scale and efficiency that algorithms bring to IOs (Recorded Future 2024).

In common with LLMs that generate human-like written text, machine learning (ML) algorithms have produced generative models that produce fake audiovisual content via deepfake technology. Such models can synthesize ultra-realistic imitations of human faces, use voice cloning, and create realistic, high-quality audio and video impersonations (Juršėnas et al. 2021, 12–16). Fake images are generated via variational autoencoders (VAEs), like the DALL-E created by OpenAI, and the applications that swap faces in photos and videos via generative adversarial networks (GANs), like the StyleGAN2. The fact that such generative models are publicly available raises serious concerns about the capabilities accessible to malicious actors. An agent of influence can use a fake image in a social media profile and mislead others into believing that the profile is authentic. DALL-E can generate images from semantic textual descriptions without requiring fine-tuning. Thus, it can be exploited to create fake images that support influence campaigns. For instance, the model can produce several images of the same scene with subtle variations, such as different angles, which enhance the perceived authenticity of fake images (Juršėnas et al. 2021, 13–14). Fake images and deepfake videos have also been utilized by Israel and Hamas. An example is an image of a burned child corpse shared by Israeli Prime Minister Benjamin Netanyahu, highlighting the atrocities done by Hamas. The picture received millions of views, but was quickly debunked as being AI-modified (Cherry 2024, 27). Likewise, on October 31, 2023, the Verify-Sy platform published a report confirming that video clips purportedly showing Israeli army tanks targeting the Palestinian resistance were actually edited footage from the video game Arma 3 (Al-Kbat 2023).

Deep learning algorithms also enable the creation of audio deepfakes for malicious purposes. Audio deepfakes are particularly concerning because they mimic biometrics and can exploit speech-based identity verification systems. They rely on text-to-speech synthesis or voice conversion systems. Models such as Tacotron, Wavenet, and DeepVoice3 are trained on real recordings to produce more natural-sounding speech, and it is even possible to create a realistic voice clone of a specific individual. A current limitation of text-to-speech synthesis is the need for detailed audio transcriptions with timestamps and annotations for nonverbal sounds to train such a model. Preparing audio transcriptions, including tasks like annotating and denoising, is time-consuming. In contrast, voice conversion systems modify a source speaker's audio to resemble a target speaker's voice while preserving the original linguistic content. This speech-to-speech conversion offers more flexibility than text-to-speech because the source speaker can control their intonation. There are plenty of open-source services, such as Google Cloud TTS, Amazon AWS Polly, Baidu TTS, Overdub, and iSpeech, offering text-to-speech and voice cloning capabilities. These AI-powered tools aim to accurately replicate key characteristics of a genuine voice, including expressiveness, roughness, breathiness, stress, and emotion. Synthetic audio can already simulate these features, making it difficult for listeners to distinguish between fake and real speech, especially in low-quality channels like phone calls (Juršėnas et al. 2021, 14–15).

AI-powered tools, such as sock puppets, cyborgs, and bots, are commonly employed in IOs, to increase the speed and range of dissemination. Sock puppet accounts are fake profiles managed by real individuals in large numbers, designed to deceive while remaining disconnected from any genuine identity. Social bots, in particular, offer a cost-effective option for malicious actors to distribute large volumes of disinformation and manipulate public perception on disruptive topics (Juršėnas et al. 2021, 17). A common practice used in IOs is that of astroturfing. The latter refers to a deceptive technique, where bots mimic authentic public opinion by flooding social media, comment sections, or other public forums with messages that appear to be coming from ordinary people. This can be done through fake accounts, scripted messages, or coordinated campaigns (Keller et al. 2020).

Future Trends and Limitations

The above analysis vividly demonstrated that the application of AI-powered tools in IOs, offers increased quantity, quality, and personalization of generated content, but also blurs the line between machine-generated and human-generated content. Bearing in mind that we are on the eve of the AI era, it is crucial to identify at this early phase, the future trends and limitations that AI brings to the art of influence. The parameters that are expected to shape the way AI-powered tools amplify or restrain the conduct and effectiveness of influence campaigns are technological, conceptual, and political.

Regarding technology, it is only natural to expect that as AI models become more advanced in the near future, this will also have an impact on the art of exerting influence. One area that will greatly benefit from improved AI-powered tools is machine translation. The latter enables the widespread reuse of disinformation across multiple languages. This capability is particularly effective for global languages with substantial speaker populations, which have contributed vast amounts of text for training automatic translation models. Ongoing advancements in machine translation will further enhance translation accuracy for both widely spoken and less common languages, thereby amplifying the dissemination of reused disinformation content (Juršėnas et al. 2021, 21).

On the other hand, AI is currently employed to combat IOs. This involves identifying social bots, screening content to detect potential disinformation, conducting in-depth analyses to uncover altered versions of previously debunked articles, tracking hostile narratives, and detecting AI-generated content such as text, images, and audio (Smith et al. 2021). One example is Data Robot, a military tool designed to scan social media platforms for misleading content, providing commanders with accurate information to support sound decision-making (Demarest and Moore-Carrillo 2023). Another is DARPA's Media Forensics (MediFor) program, which develops detection and fusion algorithms to identify manipulated images and videos. MediFor generates a quantitative integrity score for each piece of media, helping to filter and prioritize content on a large scale. A low integrity score indicates potential manipulation, flagging the media for review by analysts, allowing AI to process vast amounts of data while analysts focus on the most critical cases (Hunter et al. 2024, 14).

Actually, many of the limitations in implementing AI for IOs help to facilitate detection. When it comes to LLMs, while they offer clear advantages for generating large volumes of low-cost, low-quality content, especially in multiple languages, their practical implementation faces notable challenges. The combination of human operators and LLMs technology may create new opportunities for manipulation, but the technology alone doesn't serve as a silver bullet. In some instances, the content produced by LLMs reveals its own artificial nature, rather than being exposed by advanced detection methods (Fredheim and Pamment 2024).

Current AI models require high-quality, unbiased datasets to learn complex tasks effectively. However, creating new ML datasets is both expensive and challenging. When AI models are trained on biased or overly specific data, they struggle to generalize, limiting the effectiveness of pre-trained models that lack fine-tuning in practical scenarios. Additionally, the ongoing arms race between influence actors and defenders, such as analysts and fact-checkers, makes it difficult to keep datasets and models current. Training data must continuously adapt to the evolving information environment. One of the most critical limitations of present AI models is their lack of common-sense reasoning, which hampers their ability to fact-check, assess logical consistency, and interpret indirect statements like metaphors (Juršėnas et al. 2022, 27).

Another point to consider regarding the present and future capabilities offered by AI is that although language models, deepfake technology, and bots can increase the volume and dissemination of generated content, one must consider whether there is a need for more (dis)information. Even before the advent of AI-powered tools, a plethora of information content, both true and fake, flooded the information environment and went unnoticed. So, although AI can enhance the quantity, quality, and personalization of content, the effectiveness of an influence campaign, regardless of the technological means used, depends on gaining the attention of a targeted audience (Simon et al. 2024). Thus, future development in the area of AI-generated content for IOs should explore why and how targeted audiences reject high-quality information and are more prone to receiving disinformation.

The last factor that will shape the conduct of AI-powered IOs is policy regulations. Governments can require companies developing or deploying AI systems to provide transparency reports that detail how these systems are used to generate or amplify content. Likewise, governments should enforce AI providers to develop more detectable outputs such as digital watermarks or data fingering measures and impose stricter usage restrictions on their models. Another option is the promotion of media literacy and public awareness campaigns. Governments should invest in media literary campaigns in schools, universities, and the general public and sponsor fact-checking initiatives. At present, it is safe to argue that AI-powered tools that augment the impact of influence campaigns develop faster than countermeasures.

Conclusion

The rise of AI-powered technologies that transform the conduct of IOs is a reality. The fact that the current society integrates increasingly deeper into an information-intensive space, leads to a publicly available abundance of data that can be used for persuasion, coercion, and manipulation. Algorithms can analyze public sentiment, coordinate profiling, and target with great efficiency by knowing when a targeted audience is more susceptible to influence. The ongoing conflicts in Europe and the Middle East demonstrate that the use of deepfakes and chatbots is an emerging trend. In the near future, using such capabilities to generate fake content and amplify its reach will become the new normal. This raises significant challenges not only for states and their security apparatus, but also for the private sector and the scientific community. These challenges range from doctrinal development and operational guidelines for the armed forces to counter AI-enabled IOs, to legal frameworks and technological standards for the private sector, and the research community.

Finally, it is important to note the following points regarding the interaction between AI and IOs. First, overestimating the potential impact of AI in influence campaigns can be counterproductive. Highlighting the possibility that any online account could be an AI-powered bot might encourage people to dismiss arguments they disagree with as inauthentic, fueling further division and polarization in society. Similarly, overstating the prevalence and risks of online disinformation could inadvertently advance the very goal of the influencer – fostering a deep distrust of any information that challenges individuals' preexisting beliefs. Second, influence is a broad and hard-to-define concept that synthesizes communication studies, cognitive science, and political studies and is much more than an algorithm that generates content. The most challenging task of any influence campaign is how to make "truth" ambiguous by introducing fake or misleading facts into an information system and not whether a text is human- or machine-generated.

References

Al-Kbat, Sarah. 2023. "Gaza and the War against Disinformation." International Council Supporting Fair Trial and Human Rights, December 5. www.icsft.net/18042/

Armistead, Leigh. 2010. *Information Operations Matters. Best Practices.* Potomac Books.

Briant, Emma and Vian Bakir, eds. 2024. *Routledge Handbook of the Influence Industry.* Routledge.

Buchanan, Ben et al. 2021. "Truth, Lies and Automation. How Language Models Could Change Disinformation." *Georgetown's Center for Security and Emerging Technology.* https://cset.georgetown.edu/publication/truth-lies-and-automation/.

Cherry, Sarah. 2024. "Modern Armed Conflicts: Disinformation Campaigns Shaping the Digital Information Landscape." *The Serials Librarian,* 85 (1–4): 19–31.

Cristiano, Fabio and Bibi van den Berg. 2023. *Hybridity, Conflict and the Global Politics of Cybersecurity.* Rowman & Littlefield.

Demarest, Colin and Jaime Moore-Carrillo. 2023. "US Military Targets Deepfakes, Misinformation with AI-Powered Tool." C4ISRNET, August 1, 2023. www.c4isrnet.com/information-warfare/2023/08/01/us-military-targets-deepfakes-misinformation-with-ai-powered-tool/.

DiResta, Renée et al. 2021. "In-House Vs. Outsourced Trolls: How Digital Mercenaries Shape State Influence Strategies." *Political Communication*, 39 (2): 222–253.

Farooq, Nusrat. 2024. "Content Warfare: Combating Generative AI Influence Operations." *Tech Policy Press*, July 12. www.techpolicy.press/content-warfare-combating-generative-ai-influence-operations/.

François, Camille. 2019. *Actors, Behaviors, Content: A Disinformation ABC Highlighting Three Vectors of Viral Deception to Guide Industry & Regulatory Responses.* Transatlantic High Level Working Group on Content Moderation Online and Freedom of Expression, September. www.ivir.nl/publicaties/download/ABC_Framework_2019_Sept_2019.pdf.

Fredheim, Rolf and James Pamment. 2024. "Assessing the Risks and Opportunities Posed by AI-Enhanced Influence Operations on Social Media." *Place Branding and Public Diplomacy*. https://link.springer.com/article/10.1057/s41254-023-00322-5.

French, Laura. 2024. "Open AI Report Reveals Threat Actors Using GhatGPT in Influence Operations." *Social Media Magazine*, May 31. www.scmagazine.com/news/openai-report-reveals-threat-actors-using-chatgpt-in-influence-operations.

Fridman, Ofer, Vitaly Kabernik, and Francesca Granelli, eds. 2022. *Info Ops: From World War I to the Twitter Era.* Lynne Rienner Publishers.

Goldstein, Josh et al. 2023. "Generative Language Models and Automated Influence Operations: Emerging Threats and Potential Mitigations." *Georgetown's Center for Security and Emerging Technology, OpenAI, Stanford Internet Observatory*, January. https://arxiv.org/abs/2301.04246.

Goldstein, Josh and Girish Sastry. 2023. "The Coming Age of AI-Powered Propaganda. How to Defend Against Supercharged Disinformation." Foreign Affairs, April 7. www.foreignaffairs.com/united-states/coming-age-ai-powered-propaganda.

Gregor, Miloš and Petra Mlejnková. Eds. 2021. *Challenging Online Propaganda and Disinformation in the 21st Century.* Palgrave Macmillan.

Hunter, Lance et al. 2024. "Artificial Intelligence and Information Warfare in Major Power States: How the US, China, and Russia Are Using Artificial Intelligence in Their Information Warfare and Influence Operations." *Defense & Security Analysis*, 40 (2): 1–35.

IISS. 2023. "Large Language Models: Fast Proliferation and Budding International Competition." *Strategic Comments*, 29 (6). www.iiss.org/publications/strategic-comments/2023/large-language-models-fast-proliferation-and-budding-international-competition/.

Juršėnas, Alfonsas et al. 2021. "The Double-Edged Sword of AI: Enabler of Disinformation." NATO Strategic Communications Centre of Excellence. https://stratcomcoe.org/publications/the-double-edged-sword-of-ai-enabler-of-disinformation/221.

Juršėnas, Alfonsas et al. 2022. "The Role of AI in the Battle Against Disinformation." NATO Strategic Communications Centre of Excellence. https://stratcomcoe.org/publications/the-role-of-ai-in-the-battle-against-disinformation/238.

Karalis, Magdalene. 2024. "Fake Leads, Defamation and Destabilization: How Online Disinformation Continues to Impact Russia's iNvasion of Ukraine." *Intelligence and National Security*, 39 (3): 515–524.

Keller, Franziska et al. 2020. "Political Astroturfing on Twitter: How to Coordinate a Disinformation Campaign." *Political Communication*, 37 (2): 256–280.

Liaropoulos, Andrew. 2023. "Victory and Virality: War in the Age of Social Media." *Georgetown Journal of International Affairs*, 24 (2): 198–203.

Lin, Herbert and Jaclyn Kerr. 2021. "On Cyber-Enabled Information Warfare and Information Operations." In *The Oxford Handbook of Cybersecurity*, edited by P. Cornish. Oxford University Press. https://academic.oup.com/edited-volume/41360/chapter-abstract/352561538?redirectedFrom=fulltext.

Mazarr, Michael et al. 2019. *The Emerging Risk of Virtual Societal Warfare: Social Manipulation in a Changing Information Environment.* RAND Corporation.

Mazzucchi, Nicolas. 2022. "AI-Based Technologies in Hybrid Conflict: The Future of Influence Operations." *Hybrid CoE Paper*, 14. www.hybridcoe.fi/publications/hybrid-coe-paper-14-ai-based-technologies-in-hybrid-conflict-the-future-of-influence-operations/.

Open AI. 2024. "AI and Covert Influence Operations: Latest Trends." https://downloads.ctfassets.net/kftzwdyauwt9/5IMxzTmUclSOAcWUXbkVrK/3cfab518e6b10789ab8843bcca18b633/Threat_Intel_Report.pdf.

Ramluckan, Trishana and James Bret van Niekerk. 2019. "International Humanitarian Law and Cyber-Influence Operations." *Journal of Information Warfare*, 18 (3): 67–82.

Recorded Future, 2023. "Obfuscation and AI Content in the Russian Influence Network "Doppelgänger" Signals Evolving Tactics." *Insikt Group*, December 5. www.recordedfuture.com/research/russian-influence-network-doppelgangers-ai-content-tactics.

Recorded Future, 2024. "Russia-Linked CopyCop Uses LLMs to Weaponize Influence Content at Scale." *Insikt Group,* May 9. www.recordedfuture.com/research/russia-linked-copycop-uses-llms-to-weaponize-influence-content-at-scale.

Reuters Fact Check. 2023. "Fact Check: Fake BBC Clip on Ukrainian Politician Selling Arms to Hamas." *Reuters.com*. December 12. www.reuters.com/fact-check/fake-bbc-clip-ukrainian-politician-selling-arms-hamas-2023-12-12/.

Sedova, Katerina et al. 2021. *AI and the Future of Disinformation Campaigns: Part 2: A Threat Model*. Center for Security and Emerging Technology. https://cset.georgetown.edu/publication/ai-and-the-future-of-disinformation-campaigns-2/.

Simon, Felix, Sasha Altay, and Hugo Mercier. 2024. "Misinformation Reloaded? Fears about the Impact of Generative AI on Misinformation Are Overblown." *Harvard Kennedy School Misinformation Review*, 5 (4). https://misinforeview.hks.harvard.edu/article/misinformation-reloaded-fears-about-the-impact-of-generative-ai-on-misinformation-are-overblown/.

Smith, Steven et al. 2021. "Automatic Detection of Influential Actors in Disinformation Networks." *PNAS*, 118 (4): 1–10. www.pnas.org/doi/10.1073/pnas.2011216118.

Taylor, Philip. 2003. *Munitions of the Mind: A History of Propaganda from the Ancient World to the Present Era*. Manchester University Press.

Virtual Manipulation Brief. 2024. "Hijacking Reality: The Increased Role of Generative AI in Russian Propaganda." NATO Strategic Communications Centre of Excellence. https://stratcomcoe.org/publications/virtual-manipulation-brief-20241-hijacking-reality-the-increased-role-of-generative-ai-in-russian-propaganda/307.

Wanless, Alicia and James Pamment. 2019. "How Do You Define a Problem Like Influence?" *Journal of Information Warfare*, 18 (3): 1–15.

Whyte, Christopher, Trevor Thrall and Brian Mazanec, eds. 2021. *Information Warfare in the Age of Cyber Conflict*. Routledge.

4 The Threat Posed by Commercial First-Person View (FPV) Unmanned Aerial Vehicles (UAVs) Modified by Asymmetrical Warfare Actors

Christopher Lavers

Introduction

Over 160 years, unmanned platforms have matured as a significant war-fighting technology, from the Italian War of Independence (1849), when Austria first used balloons with bombs, to current wide-scale military and civilian use. Innovation increased markedly during COVID, with modified unmanned aerial vehicle (UAVs) operated in various ways (Sahasranamam 2020). Air power, as a force multiplier, matured in World War Two (WW2) through tactical and strategic missions, and afterwards to maintain superiority was operationally transformed by innovations, including drones, to provide optimal air power. From WW2, the shift was from large-scale forces to smaller, flexible, and adaptable forces, working well with UAV-to-drones (UtDs) in new fighting environments. Distinctions between 'civilian and military' have now virtually vanished; with the 'war-fighting footprint' changing such that there are no defined battlefronts, nor distinguishable combatants. Non-state aggressors with UAV platforms hide effectively amongst civilians, the 'urban guerrillas' domain.

Civilian-Modified Threats

The recent period (2012–2022) saw the dawn of asymmetric warfare by non-state actors incorporating small, cheap, hand-launched drones, often indigenously manufactured, using simple materials with low radar cross-section (RCS), visible, and thermal signatures (Lavers 2013). Modification of commercially-off-the-shelf (COTS) UAVs by hostile actors for nefarious activities, pose an existential threat, with terrorist operations against critical infrastructure, or improvised explosive device (IED) deployment in conventional urban operations. Understanding of this emerging threat is needed, from design to operation, to develop effective countermeasures.

In 2011, John Vilasenor stated

> people can disagree on how long it will take for terrorists, insurgents and rogue groups to acquire weaponised drones guided by video straight into a target, there is no dispute that it is a question of when and not if.
>
> (ICRS 2014)

DOI: 10.4324/9781003520160-4

With unprecedented access to technology, modified UAVs make them attractive for terrorists. Terrorist past-modified 'drone failures' are no guarantee of continued failure, as early use is rarely the most effective. Paul Scharre notes, "the history of warfare revolutions in warfare have shown they are won by those who uncover the most effective ways of using new technologies, not necessarily those who invent the technology first" (quoted in Rassler 2016). Here the terms UAV includes civilian platforms possessing *latent capacity* for weapon/payload delivery, whilst UtD or 'drone,' are modified or existing military platforms.

The Changing Nature of Conflict

UAV platforms have measured appeal to unconventional forces, swarming co-operatively, concentrating decisive synchronised offensive power, combined with surprise in complex topographically constrained operations. In the second Nogorno-Karabahk war (2020) drones augmented conventional forces in a major escalation of conflict, with weaponised and non-weaponised reconnaissance flights, Israeli Orbiter-1 'loitering munitions' and Turkish *Bayraktar-2* drones, with devastating impact (Shaikh and Rumbaugh 2020). Since 2020, after 14,040 confirmed strikes, drones have demonstrated lethality (The Bureau of Investigative Journalism 2024), liquidating Western enemies across the Middle East, transcending time and borders. Whilst debate argues around whether deployment to Yemen and Pakistan is legal or ethical (Wolff 2024), they *are* increasingly used; the real questions are *how* they are used, and *how* their capacity might be constrained through design, preventing modification.

Modified-UAV Threats

Commercial and hobbyist UAVs challenge traditional concepts of security and safety, collecting data, transporting loads, delivering ordnance, and re-shaping surveillance. Commercial UAVs are typically small, using cheap and available components from electronic shops, emerging mostly from UAV and quadcopter enthusiasts, and subject to little scrutiny, carrying sensors, detectors, communications, and in exploited capacity, ordnance. Modified UAV proliferation challenges military domination, when deployed in urban environments for hostile tasks. Non-state actor deployment, in weaponised and surveillance capacities, makes 'consumer' platforms a security concern. UAVs are inherently designed for recovery, and carry lethal or non-lethal payloads, responding flexibly with reduced costs compared with traditional devices. Modified 'drones' have been deployed for: assassination attempts, hazardous material transportation, and capturing imagery over sensitive sites (UK Parliament 2019).

Rapid advances in civilian consumer UAVs have generational designs *faster* than military ones, especially in intelligent flight and manoeuvring, raising concerns on malicious deployment. Drones have been deployed by non-state groups: surveillance gathering (Daesh, Islamic State), in weaponised capacities, embedding munitions to detonate later (PKK, and Hezbollah), overwhelming defences with swarms, besides propaganda (Rossiter 2018). Daesh modified commercial and home-made UAVs to conduct reconnaissance, filming attacks against Iraqi and Russian military bases in Syria. However, allied successes in 'neutralising' Syrian and Iraqi terrorist 'hubs' may simply disperse such technologies and UtD capabilities into wider diaspora. UtD implementation isn't just about platform acquisition, it is rather a revolutionary transformation of digital products, combining sensors' inputs, with information architecture to exploit the potential of real-time UAV operations.

Any doubts about the political threat drones pose disappeared after the 2018 attempt to assassinate Venezuelan President Maduro at a parade with 2 Da-Jiang Innovations (DJI) M600s modified with 1 kg of remotely triggered C4-laden, exploding mid-air (*CNN* 2019). Another fitted with radioactive sand landed on the Japanese Prime Minister's office, flown by an individual protesting the domestic nuclear policy (*The Guardian* 2015). Unauthorised drones have captured imagery over the Eiffel Tower, White House, and military bases, with the UK reporting 37 security breaches in 2014 (Custers 2016), besides disrupting transport infrastructure. In December 2018, Gatwick Airport suffered serious disruption for 30 hours following a drone reported over the airfield, with disruption evident at other airports, from Heathrow to Warsaw. In April 2019, documents reportedly authored by activist group *Extinction Rebellion*, detailed plans to deploy drones to disrupt Heathrow operations. However, targeting transport infrastructure isn't new, Japanese group *Aum Shinrikyo* considered drone use, but eventually distributed sarin gas on Tokyo's subway. Gatwick provided timely reflection, demonstrating the extent to which economic disruption of critical national infrastructure constitutes a national security risk, with the US Department of Defense (DoD) looking to deter attacks against its power grid (Narayanan et al. 2020).

Following such airport disruptions, the UK government extended to 5 km the range around airports to which flights are prohibited, with *The Domestic Threat of Drones* Parliamentary inquiry resulting (UK Parliament 2019), looking at the UtD evolving pathway, UK vulnerability, and counter-drone effectiveness, policy, and practice. In the hands of urban guerrillas or overseas trained terrorists, UtD unconventional methods allow targeting of political goals, civilian attacks, government disruption, or assassination. Urban terrorists may employ drones to systematically inflict damage to authorities, to wear down and demoralise, sustaining operations and tactics without defending recognised operations bases, preventing conventional forces 'squaring-off' against them, in confrontations they would likely lose. Drones provide guerrilla tactical advantages with surprise (Marighella 1969), acting in swarms at speed, making it hard for ground forces to match. Addressing emerging hostile actor threats, with resources currently allocated to the UK Armed Forces, including UtD operation, will be difficult.

Biological and chemical attack risks were raised by UK leaders before the 2012 Olympics; considering it feasible, modified UAVs might deploy poison (Daily Mail 2012). Radioactive material transported into urban areas for 'dirty-bomb' attacks is considered unlikely due to accessible materials access. UAVs are being outfitted with various payloads; besides cameras and sensors, police have fitted pepper spray/teargas, with enthusiasts adding functionalities such as fireworks, claws, handguns, chainsaws, paintball, flamethrowers, and tasers. New designs offer insights into how guerrillas may innovate UtDs, with gun-firing and mounted flame-thrower operation demonstrated. One likely exploitation route is weaponisation of agricultural pesticide dispersal systems, maximising terrorism, whilst Ukraine has seen first UAV RPG deployment (2024).

Consumer UAVs are available with modular installation of lightweight electronics devices. Under Federal Aviation Administration (FAA) regulations, they can fly up to 400 ft requiring pilot control during take-off and landing. However, terrorists, have no interest going higher as the intent is to avoid radar or visual detection, exploiting surprise. Cyber threats cannot be ignored, as UAVs require on-board computing, leaving on-board and ground-based controllers vulnerable to malicious software (Maldrone attack), and GPS(positioning, navigation and timing (PNT)) navigation attack. *Spoofing* entails sending false GPS signals towards UAVs, so they are 'hijacked' to follow programmed directions, manipulated to crash, or flown to the attacker's location and modified, with the first spoofing attack by the Department of Homeland Security in 2014. Military GPS uses encryption, rendering UAVs invulnerable to spoofing, but susceptible to *jamming*, where UAV GPS receivers are overwhelmed with signals. Encryption ensures false signals aren't mistaken for genuine ones, but signals cannot get through elevated noise, so collision may result, as incidents have caused GPS loss without jammers being found (Boyle 2015). Besides navigation, there is the issue of malign payloads, requiring platform design so they cannot *easily* be modified for ordnance, omitting weapon mounting points, for example, the UN *Falco* (medium.com, 2014), limiting exploitation. 'Stopping' hardpoints or explosives being added is almost impossible. Putting elements into the design that increase radar reflectivity, on the other hand, may be harder for criminals or terrorists to remove. Combined with the low visibility of small UAVs and low radar cross section (RCS) materials, for example, glass reinforced plastic, making UAV to drone platforms very hard to detect, a key aspect of stealth. "UAVs are designed to operate without detection and shouldn't be vulnerable at a low height. This is why I think it is the most important factor in UAV design, so both stealth and avoidance from shoot-down are achieved," *Mil. Anon.*

Technology Access

Access to, and costs associated with technology modification, and actor ability to develop expertise are important. For asymmetric attackers, new technologies usually sit outside affordability and access, except via other actors; however, with commercial or 'bespoke' hobbyist systems, this is achievable (Jackson et al. 2008). Actor ability to develop expertise depends on the extent it focuses on technology; easy for nation states, less for dispersed insurgents, and harder for smaller entities such as terrorists or 'lone-wolf' attacks. However, one attack with minimal damage may achieve significant propaganda or strategic goals. UtDs afford attackers flexibility in planning, through surveillance and targeting, and may escape the scene entirely. Drones, like planes, provide speed and versatility, introducing a paradigm shift, with autonomous software-mediated 'swarms.' Here, innovation is significant as the time to field accessible new technology is less for terrorists with a 'try-and-see-approach' rather than military procurement timescales. In "Countering Irregular Activity within a Comprehensive Approach," Rear Admiral Parry, noted "hybrid warfare is conducted by irregular forces that have access to more sophisticated weapons and systems normally fielded by regular forces."

The Case for Drones

Coalition drones used against Al Qaeda had limited success degrading and temporarily disrupting their hierarchical structure, eliminating senior leaders (Bolland and Ludvigsen 2018), and with an airstrike that killed General Qassem Soleimani, head of Iran's elite Quds force at Baghdad International Airport in 2020 (BBC 2020). Attempts to assess Al Qaeda were inconclusive on whether strikes helped *or* hampered objectives, due to the aftermath (further radicalisation and political fallout from collateral damage). Systematic American drone use in Syrian and Afghan theatres (a) denigrated group hubs, and (b) forced UtD dispersal into local insurgencies, harder to defeat with current war-fighting approaches. Until recently, few considered drones as more than assisting conventional war-fighting, but after the 2020 Armenian–Azerbaijan conflict, planners and politicians are aware of drone war-winning potential, contextually in leveraged small wars. Drones provided short-term immediate Azeri gains combined with terrain conditions, but this is unlikely to be repeated effectively as countermeasures 'catch-up' with threats.

Strategy alone cannot be expected to bring down political regimes, but UtDs radically *remodel* operational engagement rules. The *nature* of airpower may not have changed with UAV adoption, but its *characteristics* have, a summation of incremental technological revolutions which exploited together are considered evolutionary. Policy-makers and military experts endorse the concept of drones revolutionising air power because they change *how* the character of war is conducted, transforming tactics, doctrines, and concepts before a new 'equilibrium' is established. UtDs reduce collateral damage when combined with precision, with multiple capabilities within a single UAV making them game-changers for terrorist implementation with limited budgets. Modified UtD systems contain other key elements, no human is on-board, removing insurgents from the 'danger-to-life' zone with operators remote to the battlefield. Civilian and military systems, designed to be recoverable, may be fielded as low-cost IED-drones and kamikaze systems. In 2017, the '*Slaughterbot*' video role-played terrorists with drone IED-warheads killing students (YouTube 2017); however, Ukrainian deployment of anti-personnel drones demonstrates this is not fiction. "Proliferation of inexpensive, small drones world-wide is the most concerning tactical development, since the rise of improvised explosive devices in Iraq," remarked Marine General Kenneth McKenzie in 2021.

Benefits of UtD Platforms

Small commercially modified COTS UtDs are cheap compared with manned aircraft, appealing in their range of applications, payload versatility and cost, and removing a key warfare restraint, namely the risk to one's own forces. UtDs make asymmetric actor conflict in Ukraine easier to conduct and operations more likely. Modified UAVs fit well with unconventional war-fighting, but lack accountability, with unclear distinction between war and not war, operated in the *grey zone* (Connable 2020). Regarding terrorism, Professor Rosa Brooks argues "there is no such time as peacetime" and the "forever war" is here. Many hostile measures: sabotage, to political destabilisation, are measures falling short of conventional war, where UtDs are potent tools in the hands of unconventional forces, revolutionising air terrorism, insurgency, or guerrilla tactics. Pan-national terrorism, crosses state boundaries with impunity, impacting entire regions. According to Evans "The possibility of continuous, sporadic, armed conflict, its engagements blurred together in time and space, waged on several fronts … , means that ... war ... is likely to transcend a neat division into distinct categories" (Evans 2003).

Modified UAV Notable Events

Lone Wolf: In 2011, a 'Lone Wolf attack' was intercepted by FBI agents, planning to fly explosives-laden aeroplanes into the Pentagon and US Capitol with mobiles phones to detonate IEDs. In 2015, a National Geospatial-Intelligence Agency employee lost control of a DJI Phantom quadcopter, which crashed onto the White House Lawn. Four months later, another man was arrested flying a *Parrot Bebop* drone over the White House fence.

Terrorist organisations: In 2006, Hezbollah launched *Ababil* drones, carrying explosives, to attack targets, but were shot down by Israeli F-16s. In 2012, Hezbollah allegedly flew a small *Ayub* drone 50 km into Israeli airspace, undertaking reconnaissance on a nuclear reactor, or the presumed Kataib Hezbollah drone attack in Jordan (January 2021) resulting in the first case of US soldiers killed by drone. In 2014, during Operation Protective Edge, Israel shot down a Hamas-control *Ababil-1* (Al Qassam Brigades), with a *Patriot* surface-to-air missile; whilst Al-Qassam captured an Israeli *Skylark*-1 over Gaza, claiming it was repurposed and operational (2015). To Hamas, drone size was irrelevant, achieving subversion of technological dominance, kudos, and victory over a high-tech 'Goliath' (Blount and Sammut 2023). The group claimed, "this is a great achievement … and a gift to the Palestinian people demonstrating the strength of our people and its resistance."

An IS-modified DJI *Phantom* platform in 2014 showed Fallujah, Iraqi propaganda imagery, (Iraq 2014), providing Intelligence, Surveillance and Reconnaissance (ISR) and target acquisition, and battlefield videos during the Baiji oil refinery assault (Weiss 2015). Islamic terrorist threats aren't simply a Middle East issue, *Boko Haram* in Nigeria regularly uses drones against government forces in the North East (2018–2025) (Ogundipe 2018). US Yemeni strikes were reciprocated by terrorist groups, with Houthis killing 32 military personnel at a public parade in August 2019 (YouTube 2019).

Insurgents: Just prior to the Russian invasion of Ukraine (February 2022), Donetsk People's Republic's militias in eastern Ukraine reportedly deployed Russian-made *Eleron*-3SV drones modifying hobbyist UAVs for ISR support, and militias experimented with GPS spoofing and signal jamming against Ukrainian drones.

Potential Applications

Surveillance supports different activities from ISR, critical in the fight against *piracy and terrorism* in countries with substantial coastlines in the Gulf of Guinea, and Horn of Africa, with high incidence threatening international shipping, especially KSA, with recent Iranian and Yemeni operations. Ships and seaports (port security) are vulnerable, with potential for *critical infrastructure* attacks. Threats to oil refineries (e.g. Aramco Abqaiq attack, September 2029) and offshore operations (oil platforms) involves intruders with surface and air kamikaze craft, or small submarines, with kamikaze surface autonomous boat swarms used extensively in Ukrainian waters.

UtD technical characteristics sit at the heart of operational performance. Endurance is the main parameter measuring how long UAVs stay persistently aloft, varying between designs, *ScanEagle* has >24-hour endurance, whilst the AeroVironment *Puma*, has 3.5-hours endurance. Most fixed-wing small COTS-modified UAVs have endurance 60–90 minutes with small military ones similar to civilian ones. Endurance may increase with modification, sacrificing payload for batteries. FPV data-link range is typically 1–10 km. Maximum altitude is constrained by operational choice, and payload constrained by maximum range or sensor field of view (FOV), with small UAVs flying up to 400m altitude. Hobbyists or commercial operators are constrained by regulation to avoid air traffic, but non-state actors don't respect regulations, unless keeping to flight paths supports their objective, for example, appearing as commercial UAVs until the last moment. Speed varies with weight, design, and propulsion, with 10–50 m/s for small fixed-wing UAVs with high speed possible with pulse-jet engines.

Cameras are the main payloads carried by small FPV UtDs, for piloting or secondary objectives. Cameras and payloads depend upon: application, size, and budget. The simplest and lowest-cost video systems use analogue rather than digital cameras, sending video to ground via amateur radio-links.

Endurance is key for surveillance, alongside range and mass, "the greater the endurance the longer the range and the more time UAVs can conduct recon or support with its payload." With regard to small, modified drones: "The biggest problem we faced is flight endurance, drone use was 20 minutes, adding limitations and difficulties we faced in Qarou island, and Failaka island."

Propulsion: One option is the pulse jet, similar to unmanned vengeance weapons, now used by hobbyists world-wide, providing high-speed flight which non-state actors may want to mount attacks where speed provides tactical advantage.

Capacity: Payloads are installed near the nose, or fuselage, allowing clear FOV. Maximum capacity depends on design, but generally 10–20% of gross take-off weight (GTOW), regardless of application. High-end commercial UAVs are equipped with cheap cameras, thermal imagers, loudspeakers, and spotlights, ideal for UtDs. The main civilian supplier, DJI (China) accounts for 50% of the US market, dominating >$500 USD; below this there are competing companies providing modifiable platforms.

Civilian Classification

Fixed Wing: It provides lift from fixed aerodynamic surfaces, whilst *multi-rotor* achieves lift through rotation, at higher cost and lower endurance and speed, making them less popular for hostile actors. Frames are important, light enough to take off yet providing support to survive minor crashes. There are many frames, from the common Quadcopter, to Hexacopters for heavier loads.

Very low-cost close-range UAVs have 5 km range, 45-minutes endurance, costing under US $10k. UAVs in this class are similar to model airplanes. *Close-range:* up to 50 km and 6-hours endurance, ideal for reconnaissance and surveillance.

Very-small UAVs have dimensions up to 50 cm length with flapping/rotary wings, small, light-weight, ideal for spying. 'Flapping' allows perching on surfaces, for example, Israel's *Malat* Mosquito (35 cm wingspan and 40-minutes endurance, or US *Aurora'* (60 cm wingspan, 33 cm length). Larger ones use conventional aircraft configurations.

Small UAVs have one dimension over 50 cm but under 2 m, with designs based on fixed-wing models, and mostly hand-launched, for example, *RQ-11 Raven.*

Medium UAVs are too heavy for one person to carry and smaller than light aircraft, with payloads: 100–200 kg, like the Hunter, but beyond the limits of insurgent operations unless supplied by state actors.

Range Limitation

An important problem is small drones and modified UtDs are designed for short range, which is ideal for one-way missions, unlike conventional armed drones which are generally easier to shoot down, less attractive to non-state actors than small 'stealthy' UtDs.

Long-range, low-technology: Iranian *Ababil*'s use basic remote-control to achieve flight and video recording, a technology available in electronics shops world-wide, including high-resolution cameras. These are easily shot down, and because RF-links are vulnerable to jamming or interception, aren't regarded as a major threat to the USA.

Short-range, high-technology: Many American and overseas companies have developed low-cost *systems,* with poor situational awareness. Technology is inherently dual-use using commercial inertial navigation units, found in toy helicopters and Wii controllers, ideal for indigenous modification. US doctrine for short-range air defence is mostly concerned with defeating attack helicopters with missiles rather than UAV attacks. The UK should develop defensive systems as the threat from small UtDs grows, especially short-range UtDs. Small, persistent aircraft operating autonomously are ideal for '*swarm use*' and although high-end small UAVs are vulnerable to ground fire, they are harder to target, due to low visibility, and expendable.

Short-range, low-technology: Radio-controlled airplanes have been available for decades, modified they may deliver small payloads to sensitive sites, but a successful UK attack hasn't yet been achieved, but this doesn't guarantee continued failure. Cheap GPS improves UAV ability to accurately find targets, and this category represents the most immediate threat.

Reusable vs. Expendable?

Otherwise, reusable drones conduct one-way '*kamikaze*' missions, as launching UtDs to crash into targets is easier than launching them and landing them safely. When survivability isn't an issue, small systems may launch from covert locations; weak state and sub-state groups have done this. Hezbollah used them against Israel in several unsuccessful attacks; in April 2013, the Israel Defense Forces (IDF) detected a drone on radar and dispatched F-16 fighters to destroy it. Hezbollah's poor UAV success rate contrasts with the effectiveness of ballistic missiles fired from Gaza. Israel's expensive success against expendable, low-technology UtDs results from their vigilant state, with pro-active engagement rules. In the more-relaxed USA airspace, it is possible terrorists might launch expendable armed attacks from Canada, Mexico, or from within.

Technologies making reusable UAVs such as *Predators* attractive are largely irrelevant in expendable applications. GPS navigation makes it easier for expendable UtDs to hit specified targets—particularly surprise attacks in which defences are unprepared without precautions like GPS jamming. For terrorist WMD use, precise target location is unimportant. Delivering munitions to random locations within cities may be sufficient to satisfy terrorist objectives. Drones are preferable for terrorist missions for overwhelming reasons: they are small and cheap, and no hostile actors are at direct risk of injury. For terrorists in the field, for practical logistics, there is limited interest in UAVs like Reapers. If the platform is large, the advantage of being unmanned diminishes, it is harder to set up and launch, and requires increased manpower.

Civilian Air Defence Implications

Civil Aviation Authorities track aircraft at other than very low altitude, to prevent collisions amongst civil aircraft. If a Predator loitered 15,000 feet in Mexican airspace, ATC would know—and if the Predator hadn't filed a proper flight plan, the Air Force could shoot it down. However, technologically advanced actors might hijack civilian UAVs with non-kinetic technologies such as jamming or spoofing, or small UtDs to attack aircraft near international airports.

UAV Design

Usual platform design factors, robustness, strength, redundancy, and simplicity, are less critical for short-duration, and kamikaze missions, whilst fatigue and corrosion, for example, are long-term issues (Suresh 1998).

Hobbyist Materials

Airframe materials include aluminium, heavier than carbon fibre with some home-made platforms using wood. Generally, materials have low visibility, are small, with low RCS, ideal as 'stealthy' UtD platforms, coupled with low thermal signature. As seen in Ukraine, 'cardboard' IED-drones are effective; others include printed circuit board, similar to fibreglass, thermoplastics, and epoxy-laminate G10, cheaper than carbon fibre, but more expensive than wood, or plastic.

Safe navigation is critical, hence concerns about communications security, as the 'Return-To-Origin' (RTO), command could be hacked. In 2019, Cama exploited the infotainment browser in a Tesla Model-3, inputting code and taking system control (Ornes 2020), with similar attack opportunities existing on other platforms, or to establish critical locations placement. From the RTO attack perspective, if a fault is detected, this may not be best policy: "it may be best to go to preset location to avoid compromising pilot location."

Chemical, Biological, Radiological, or Nuclear (CBRN)

"The prospects of politically violent non-state actors utilising CBRN weapons has captured public imagination" (Asal 2012). However, materials access generally precludes CBRN capability. Technical and knowledge-sharing innovations make it *potentially* easier to achieve CBRN capability, facilitating pursuit of materials for UtD delivery. However, for terrorist groups it is "comparatively unlikely to get its hands on anything truly devastating. Groups harbouring such goals would probably be small and poorly resourced .... Such WMD attacks as they are able to mount, would be a far more modest 'homebrewed' kind and therefore within limits of containability" (Abbott et al. 2016).

However, deadly IED UtDs are the likeliest demonstrated home-made urban threat, a combination of 'everyman' drone and mobile IED. The knowledge, resources, and technical skills to produce them is freely available. According to Cohen, FPV drones carrying home-made bombs to produce a new level of domestic terror are 'just around the corner,' without accessibility issues posed by CBRN (Schmidt and Cohen 2013). There are better and simpler ways than CBRN to deliver potent payloads to target. Mass is a key factor for hostile actors, it determines the potential attack payload. "Drones are technologically limited as to the weight they carry, and generally aren't capable of highly sophisticated attacks by terrorists unless assisted by very capable actors, probably a nation state," according to Walker-Roberts (UK Parliament 2019, DTD0003). "The modified UtD payload is its primary function, whether that is sensors and imaging, or bombs/missiles." Consequently, commercial UAVs should be designed to fulfil their primary role, providing little spare capacity to minimise exploitation.

Cyber security: securing against terrorist cyber-attack is vital. In the UK, the ASTRAEA program is securing code so no-one can tamper or hack it. However, "Man-in-the-middle attacks from skilled hackers, can remotely down or hijack drones. Designing UAVs that cannot be used for nefarious purposes is harder to achieve," ... and may be bypassed.

Further professional views state, 2Platforms must be licensed and legally approved to avoid unethical usage such as terrorism. But will terrorists follow licensing actions? Asymmetric actors may modify UAVs to target individuals, kill, destroy, or conduct activities to disrupt a state, impacting economy, or damaging psychological/morale of civilian populations as seen in the recent Israel-Gaza conflict after October 7, 2023. Unfortunately, "Dual-use cannot be legislated against. How do you prevent lines of code from a civilian coder being used in military items?" This inability to prevent post-sale physical or digital adaptation leaves only pre-sale legislative action. "Unfortunately, militarisation of civilian engineered products is inevitable. Legal enforced frameworks must be put in place, … allowing easier detection of criminal development outside of this framework."

Fatic's comprehensive study on traditional offensive military drone ethical framework *The Ethics of Drone Warfare*, suggests new technologies radically changes the military values system so most traditional ethics are inapplicable (Fatić 2017). Unsurprisingly, rogue states and non-state actors are unconstrained by Armed Conflict laws, gaining tactical advantages that cannot be reciprocated by law-abiding nations.

Justice appears to have no role in drone operations, except killings of 'the wanted dead or alive' category. What is the role of traditional virtues such as 'respect the enemy,' with operators hundreds of kilometres away? General Soleimani's execution in 2020 was another step in sub-war threshold West-Islamic encounters. The US–Iran 'drone war' started with the US shooting down an Iranian drone over Iraq in 2009. In 2011, Iran shot down a US RQ-170 on an Afghan mission. In 2017, US fighters shot down two Shahel-129s in Syria, and in 2017 accused Iran of supplying drones to Houthi rebels in Syria, with Iran, in turn, claiming to have eliminated a Global Hawk (Gulf of Oman), with the US downing another Iranian drone over the Persian Gulf (2019).

General Soleimani's death took the US–Iranian drone war to new levels, and may prove counter-productive to long-term peace efforts. Soleimani wasn't an Iraqi insurgent, nor terrorist, but General of a foreign national force. Legally what crime was committed at the point of execution, and what of civilian collateral deaths and damage? Incorrect target designation of civilians does occur, with one mistaken Predator transcript having eliminated a civilian convoy of Afghan women and children. AI-trained target recognition may help, but terrorist-trained software could also, for example, 'recognise' a US President.

Potential military target attacks: "It is still a relatively new area, and people tend to be reactive – hence there is a lot of awareness of drone danger where attacks have taken place e.g. Middle East, but not domestically in the UK."

Media influences public perception about military drone use, creating uncertainty about ethics and efficiency. There is a lack of clarity on *how* international law and regulatory frameworks applies to military use, with legal definitions blurred by 'grey zone' operations, and terms such as *unlawful combatants*, and *pre-active* targeted lethal action unclear. Threat assessment of risk probabilities, and damage, are essential before a major public event occurs. Response mustn't be driven by ill-informed politicians chasing public clamour, activist groups, and not a poorly considered response after an incident. Hence, its importance now, before the 'horse bolts.'

Terrorism

UtDs may be used effectively by terrorists in the littoral offensively, attacking shore-side targets (power, industrial, desalination), maritime vessels, off-shore platforms, port infrastructure, 'hub ports,' for example, Rotterdam, global choke points, for example, the Strait of Hormuz, or for propaganda/strategic purposes. As insurgents use drones more, they may increase credibility with indigenous populations, using them as effective political tools. Poplin states "militant groups are eager to celebrate legitimacy in a propaganda war that has taken on increased importance" (Poplin 2014). However, insurgent actors won't likely achieve sustained competitive advantages in modified-UAVs but rely on them more. Terrorist use *may* vilify technology such that the legitimacy of *our* drone use is questioned. Presently, maritime terrorist activities are logistically dependent, requiring movement of insurgents, arms/equipment, money/documents, a weakness in their supply chain. Future scenarios include suicide bombs, with small fast UAV swarms, likely in 'hit-and-run' attacks on land, sea, or urban, using small platforms firing assault rifles, or RPGs. Remote or autonomous control permits stand-off attacks to drop bombs or mines cheaply and effectively, placing explosive devices in ports, ships, or well-guarded buildings, combined with a Cyber assault on port-services and infrastructure, overwhelming defences.

Conclusion

Current small commercial UAVs, are easily modified and armed, overlapping with military drones sufficiently to warrant concern. Cheap hobbyist drones, purchased by individuals or groups with lower sophistication compared with higher-end military systems, require no significant infrastructure to operate, and in uncontested airspace, disrupt transport, creating fear, or carry ordnance.

The potential for neutralising hostile actor drone 'forces' on the ground, prior to launch, is minimal. They are hard to detect, covert, and many hand-launched, providing little exposure to air/space surveillance, and easily inserted within urban environments. It may become harder to attribute the source of attacks, especially if civilian UAVs are 'hijacked.' Swarming is demonstrated for mass-precision attacks which may overwhelm defences due to numbers, especially in poorly defended civilian areas. As autonomy and multi-vehicle swarm control matures, we must anticipate inexpensive, expendable UtDs against ships, shore-based infrastructure, military and civilian targets, at home and abroad in future conflict. Defence against UtDs should be high priority. At sea platforms with greater range, endurance, and payload capacity, may launch small boats and flying swarms, or mount piracy attacks.

Whilst terrorism isn't new, there is growing realisation since 9-11 of the dangers over land and sea. Existing domestic and international legal measures are inadequate against lone-wolf attacks, or groups not keeping to the 'letter-of-the-law'; since 2010, these have been increasingly deadly. The 'bomb-in-the-box' concept, small containers to smuggle/detonate CBRN in a high-priority critical infrastructure location, or one-off attempt in public spaces, although unlikely, have symbolic value. UtDs releasing biological/chemical agents without warning, are a risk, and only detected when cases of unknown origin start appearing in hospitals. Modified UtD-delivered payloads, or captured military drones, will augment terrorist objectives, but basic and more reliable methods can deliver more potent payloads to targets.

The reality is exploitation cannot be stopped. "Is there some technical boundary that makes sensors exclusively military that I'm not aware of?" We should be under no illusions, if all reasonable safeguards are in place, and export restrictions effective, 'scales are tipped' in favour of what *should* limit home-made modification, but as seen in Armenia and Ukraine, there are always nations prepared to unilaterally supply arms to one side of a conflict.

Small COTS platforms, embracing emerging and sophisticated technological capabilities can overcome human limitations encountered in conflicts. Strategy and tactics are largely defined and updated by technology, and UAVs are no exception, entering the terrorist's lexicon, and profoundly altering the way contemporary war is organised and conducted. There are no simple safeguards to deal with small modified UtDs and drones, designed agnostic to intent, but operated in a hostile capacity, are definitely here to stay.

References

Abbott, Chris, Steve Hathorn, and Scott Hickie, 2016, January, *The Hostile Use of Drones by Non-State Actors Against British Targets. Oxford: Remote Control Project, Oxford Research Group.*

Asal, Victor H., 2012, "Connections Can Be Toxic: Terrorist Organizational Factors and the Pursuit of CBRN Weapons." *Studies in Conflict & Terrorism* 35, no. 3: 229–254.

BBC News, 2020, "Qasem Soleimani: US Kills Top Iranian General in Baghdad Air Strike." January 3, 2020. www.bbc.co.uk/news/world-middle-east-50979463.

Blount, Clive, and C. Sammut, 2023, "'A Gift to Our People': The Use of Drone Technology by Islamist Insurgents." *Air Power Review* 19, no. 1: 1–24.

Bolland, Tom, and Jørgen A.L. Ludvigsen, 2018, "'No Boots on the Ground': The Effectiveness of US Drones Against Al Qaeda in the Arabian Peninsula." *Defense & Security Analysis* 34, no. 2: 127–141.

Boyle, Michael J., 2015, "The Race for Drones." *Orbis* 59, no. 1: 76–94.

CNN, 2019, "Dentro del complot de agosto para matar a Maduro con drones." *YouTube Video*. www.youtube.com/watch?v=VhuMy15rIVo.

Connable, Ben, 2020, "Russia's Hostile Measures: Combating Russian Gray Zone Aggression Against NATO in the Contact, Blunt, and Surge Layers of Competition." *RAND Corporation,* 2020. www.rand.org/pubs/research_reports/RR2539.html.

Custers, Bart, ed., 2016, "The Future of Drone Use: Opportunities and Threats from Ethical and Legal Perspectives." *Information Technology and Law Series*, Vol. 27. Springer.

Daily Mail, 2012, "Poison Drones: New Olympic Threat Warns Colonel in Charge of Keeping London Calm." May 6, 2012. www.dailymail.co.uk/news/article-2140173.

Evans, Michael, 2003, "From Kadesh to Kandahar." *Naval War College Review* 56, no. 3: 1–9.

Fatić, Aleksandar, 2017, "The Ethics of Drone Warfare." *Filozofija I Društvo* 28, no. 2: 349–364. https://doi.org/10.2298/FID1702349F.

The Guardian, 2015, "Drone 'containing radiation' lands on roof of Japanese PM's office, 22 April. hwww.theguardian.com/world/2015/apr/22/drone-with-radiation-sign-lands-on-roof-of-japanese-prime-ministers-office

Houthi Drone Targets Senior Yemeni Officers, Kills Five Soldiers. *YouTube Video*, 2019. www.youtube.com/watch?v=XQNCbzTn8nM.

International Committee of the Red Cross, 2014, *Report of the ICRC Meeting on Autonomous Weapon Systems, 26-28 March 2014.* November 1, 2014. www.icrc.org/en/document/report-icrc-meeting-autonomous-weapon-systems-26-28-march-2014.

Jackson, Brian A., Daniel R. Relinger, Michael J. Lostumbo, and Robert W. Button, 2008, *Evaluating Novel Threats to the Homeland.* RAND Corporation, 2008. www.jstor.org/stable/10.7249/mg626dtra.12.

Lavers, Christopher R., 2013, *Reeds Vol 14: Stealth Warship Technology.* London: Bloomsbury Press, 2013. ISBN 9781408175255.

Marighella, Carlos, 1969, *Minimanual of the Urban Guerrilla.* Abraham Guillen Press. ISBN 1894925025. Available at Marxists.org.

Narayanan, A., J.W. Welburn, B.M. Miller, S.T. Li, and A. Clark-Ginsberg, 2020, *Deterring Attacks Against the Power Grid: Two Approaches for the U.S. Department of Defense.* RAND Corporation, 2020. ISBN 978-1977404169. www.rand.org.

Ogundipe, Samuel, 2018, "Boko Haram Using Drones, Foreign Fighters Against Nigerian Troops." *Premium Times*, 2018. www.premiumtimesng.com/news/headlines/298105-boko-haram-using-drones-foreignfighters-against-nigerian-troops-buratai.html.

Ornes, Stephen, 2020, "How to Hack a Self-Driving Car." *Physics World*, August 2020.

Poplin, Cody M, 2014, "Look Who Else Has Drones: ISIS and Al Nusra." *Lawfare*, October 24, 2014. www.lawfareblog.com/look-who-else-has-drones-isis-and-al-nusra.

Rassler, Don., 2016, *Remotely Piloted Innovation: Terrorism, Drones and Supportive Technology.* West Point, NY: Combating Terrorism Center, October 2016. https://ctc.westpoint.edu/remotely-piloted-innovation-terrorism-drones-and-supportive-technology/.

Rossiter, Ash, 2018, June, "Drone Usage by Militant Groups: Exploring Variation in Adoption." *Defense & Security* 34, no. 2: 113–126.

Sahasranamam, S., 2020, *India: How Coronavirus Sparked a Wave of Innovation.* https://theconversation.com/india-howcoronavirus-sparked-a-wave-of-innovation-135715.

Schmidt, Eric, and Jared Cohen, 2013, *The New Digital Age: Reshaping the Future of People, Nations and Business*. London: John Murray.

Shaikh, Shaan, and Wes Rumbaugh, 2020, *The Air and Missile War in Nagorno-Karabakh: Lessons for the Future of Strike and Defense.* Center for Strategic and International Studies, December 8, 2020. www.csis.org/analysis/air-and-missile-war-nagorno-karabakh-lessons-future-strike-and-defense.

"Slaughterbot," 2017, "Autonomous Killer Drones." *YouTube Video*. www.youtube.com/watch?v=ecClODh4zYk.

Suresh, S., 1998, *Fatigue of Materials.* 2nd ed. Cambridge: Cambridge University Press, 1998.

The Bureau of Investigative Journalism, 2024. "Drone War." Accessed February 4, 2024. www.thebureauinvestigates.com/projects/drone-war.

UK Parliament. *Domestic Threat of Drones Inquiry*. October 2019. https://committees.parliament.uk/work/2150/domestic-threat-of-drones-inquiry/publications/.

War Is Boring, 2017, "The U.N.'s Drone Air Force Has Arrived: Peacekeepers Deploy Unmanned Aircraft in Congo." *Medium*, 2014. https://medium.com/war-is-boring/the-u-n-s-drone-air-force-has-arrived-7e8189300df4.

Weiss, Caleb, 2015, "Islamic State Uses Drones to Coordinate Fighting in Baiji." *Long War Journal*, April 2015. www.longwarjournal.org/archives/2015/04/islamic-state-uses-drones-to-coordinatefighting-in-baiji.php.

Wolff, Stefan, 2024, *The Ethics of Warfare Part 3: How Does Drone Warfare Change the Debate?* University of Birmingham, 2024. www.birmingham.ac.uk/research/perspective/drones-wolff.aspx.

5 Space and the New Frontier of Warfare

Markos Trichas and Matthew Mowthorpe

Introduction

During the long years of the Cold War, and despite the development and trials of anti-satellite weapons, an unwritten agreement adhered to by the Soviets and the Americans was that their military satellites would remain off limits to attack (Harding, 2023). This culminated in the Outer Space Treaty of 1967 which prohibited the deployment of weapons of mass destruction in space. In the three decades since that conflict ended, the boundaries of space warfare have eroded to the point that in 2019, NATO formally declared space to be an operational domain (NATO, 2019).

Space assets have significantly changed not only our everyday lives but also the way we conduct warfighting. As an example, Position, Navigation and Timing data from space have transformed the transport and financial sectors that enable "just-in time" supply chains, support millions of pounds of global financial transactions and synchronization of everything from utility networks to mobile phone systems (Antrobus, 2020). Meteorological satellites have transformed the way societies plan their everyday activities, from cultivating the land, monitoring the climate/environment, improving food production to planning their work and holiday travel. Communication satellites broadcast our news and sports while mega-constellations, like Starlink, provide broadband services direct to our homes regardless of location.

Our reliance on space assets means that they have now become targets for our adversaries. On February 24, 2022, a cyber-attack against commercial communication company Viasat, began approximately 1 hour before Russia launched its major invasion of Ukraine (UK Gov, 2022). Although the primary target is believed to have been the Ukrainian military, other customers were affected, including personal and commercial internet users. Wind farms in central Europe and internet users were also affected. However, the war in Ukraine demonstrated the value of space as a game changer. The provision of Starlink enabled Ukrainian forces to communicate on the battlefield and in some instances transmit to unmanned aerial vehicles (UAVs) to target Russian forces (Trichas, 2024). In addition, commercially provided satellite imagery dramatically enhanced the performance of Ukrainian forces, leading the Russian Foreign Ministry to label commercial satellites as

DOI: 10.4324/9781003520160-5

"quasi-civilian infrastructure that may be a legitimate target for a retaliatory strike" (Reuters, 2022).

Russian Counterspace Program

Under Vladimir Putin, Russia has reinvigorated its political desire to obtain counterspace capabilities for the same reason as China, to advance its regional power and limit the ability of the US to counter Russia's freedom of action (Mowthorpe, 2022). Russian military thought sees modern warfare as a struggle over information dominance and netcentric operations that can take place without clear boundaries. Russia is pursuing the goal of incorporating EW capabilities throughout its military to both protect its own space-enabled capabilities and degrade or deny those capabilities to its adversary. In space, Russia is seeking to mitigate the superiority of US and NATO space assets by fielding a number of ground, air, and space-based offensive capabilities (Mowthorpe, 2022).

Co-Orbital Assets

Although the Soviet Union invested some effort to develop anti-satellite weapons, like the co-orbital Isrebitel Sputnikov (Weeden & Samson, 2021), Russian counterspace capabilities have significantly evolved since 2010s.

On December 25, 2013, three small satellites were launched into low Earth orbit (LEO) which looked like a routine Rodnik satcom activity. The Russian MOD publicly announced the three satellites, Cosmos 2488, 2489 and 2490, had successfully separated from the upper stage. However, a fourth payload, Cosmos 2491, was catalogued by the US military. Cosmos 2491 remained dormant until the end of 2019, in LEO at an altitude of 1500 kilometres. Cosmos 2491 was identified by NASA as a secretive Russian satellite which performed orbital rendezvous and inspection manoeuvres (NASA, 2025a).

On May 23, 2014, during another Rodnik mission, three military satellites were declared by the Russian government: Cosmos 2496, 2497 and 2498. Similar to the 2013 launch, a fourth payload was identified, Cosmos 2499. In mid-June 2014, Cosmos 2499 began a series of manoeuvres to match the orbit of the Briz-KM upper stage that had placed them in orbit. At the end of November 2014, Cosmos 2499 passed within a kilometre of the Briz-KM. They then drifted apart, until in January 2015 Cosmos 2499 did a further series of manoeuvres to achieve an orbit a few kilometres above and several hundred kilometres away from the Briz-KM. On March 26, 2016, Cosmos 2499 adjusted its orbit slowly bringing it closer to the Briz-KM by about tens of kilometres per day (Weeden & Samson, 2021).

On March 31, 2015, three Gonets-M satellites were launched and openly declared as Gonets M11-M13, along with a classified military payload, Cosmos 2504. In April 15, Cosmos 2504 maneuverer to bring it close to the Briz-KM upper stage. Between April 15 and 16, 2015, Cosmos 2504 went from an estimated 4.4 kilometres to 1.4 kilometres below the Briz-KM (Weeden, 2015). On July 3, 2015,

Cosmos 2504 lowered its apogee and perigee by around 50 kilometres each, manoeuvring away from the Briz-KM. After a period of inactivity, on March 27, 2017, Cosmos 2504 lowered its orbit, and passed within two kilometres of a piece of Chinese debris from the 2007 ASAT test. This could indicate that Cosmos 2504 was an inspection satellite (Weeden & Samson, 2021).

On June 23, 2017, Cosmos 2519 was launched, which Russian officials included "a space platform which can carry different variants of payloads" (RussianSpaceWeb 2024). It made a series of small manoeuvres in late July and August. On August 23, 2017, a small satellite designated Cosmos 2521 separated from Cosmos 2519. Cosmos 2521 was declared by Russian officials as "intended for the inspection of the condition of a Russian satellite" (RussianSpaceWeb). On October 30, Cosmos 2523 another small satellite, separated from Cosmos 2521. Cosmos 2523 was released at a velocity of 27 meters per second. At this speed, it appears likely that Cosmos 2523 could be a projectile and part of an ASAT mission. Throughout March, April and June 2018, Cosmos 2519 and 2521 conducted several RPOs of each other (Weeden & Samson, 2021).

On July 10, 2019 Russia launched another set of four military payloads, designated Cosmos 2535, 2536, 2537 and 2538. On August 7 to 19, Cosmos 2535 and 2536 began a series of RPO with approach distances as close as 30 kilometres before backing off to 180 to 400 kilometres (Mowthorpe, 2022).

On November 25, 2019, Russia launched Cosmos 2542, which was likely the second satellite in the Nivelir series. On December 6, Cosmos 2542 released a sub-satellite, Cosmos 2453, which remained within two kilometres of Cosmos 2542 for three days before it conducted a series of manoeuvres to raise its apogee to 590 kilometres by December 16. These manoeuvres suggest that Cosmos 2453 moved to where it could observe a US intelligence satellite, USA 245. Cosmos 2453 came within 20 kilometres of USA 245 several times in January 2020. This proximity sparked concerns from the then-commander of US Space Command. It is likely that Cosmos 2453 was an inspection satellite.

In June 2020, Cosmos 2543 manoeuvred to come within 60 kilometres of Cosmos 2535. On July 15, similar to the event of the first Nivelir, a small piece of debris separated from Cosmos 2543 at a relative velocity between 140 to 186 meters per second (Weeden & Samson, 2021). It is likely this is a similar event to Cosmos 2523 in October 2017, which was the first of Russia's Nivelir test program. Both the US and UK Space Commands called on Russia to desist their testing of the system (*The Guardian*, 2020).

On August 1, 2022, a Russian Soyuz 2.1v launch vehicle placed a mysterious satellite, dubbed Cosmos 2558 (2022-089A, 53323) into LEO. The launch timing and initial orbit appeared to coincide with the orbital plane of USA 326, a classified NRO imagery satellite that was launched in February 2022. Analysis suggested that the orbits of Cosmos 2558 and USA 326 were very similar in inclination and would periodically come within 60 to 70 km in altitude. On August 18, 2022, USSPACECOM released a statement condemning Russia for this behaviour, calling the activities of Cosmos 2558 "dangerous and irresponsible behaviour."

Further analysis confirmed that as of September 2022 Cosmos 2558 had altered its orbit to continue to match the orbital plane of USA 326, although it is not in an actual proximity orbit. It is unclear whether Cosmos 2558 is related to Cosmos 2535 or Cosmos 2542.

It is highly likely that Cosmos 2576 launched on May 16, 2024 is part of the Nivelir ASAT programme. It is the fourth in the series of co-orbital ASAT testing satellites, similar to those condemned by both the US and UK Space Commands previously.

Direct Ascent Assets

Besides the highly advanced co-orbital assets, Russia has been developing highly capable direct ascent systems. Direct-ascent systems are ground based systems, often mobile which include a ground to space missile designed to intercept a target satellite. Such a Russian system, named Nudol, has been tested around ten times with varying levels of success. Almaz-Antey, whose principal role is active space defence technologies, has pitched the system as valuable for holding US LEO assets at risk (Weeden & Samson, 2021). Nudol is a TEL-based system composed of the 14A042 Nudol rocket, 14P078 command and control system and 14TS031 radar. In November 2021, Russia successfully intercepted one of its own satellites in LEO, using Nudol (*The Washington Post*, 2021). The operational capability of NUDOL is up to 850 kilometres. It is likely, given successful testing, Nudol will be operational by 2025.

Electronic Warfare as Counterspace

Russia demonstrated its GPS jamming capability during the Russian 2017 Zapad military exercise (The War Zone, 2019) and during a NATO exercise, when Norway determined Russia was responsible for jamming GPS signals in the Kola Peninsula during Exercise Trident Juncture (Episkopos, 2021). The Organization for Security and Co-operation in Europe (OSCE) in April 2021 identified in Ukraine, an increase in GPS jamming by Russia or pro-Russian forces. On April 6, 2021, a Special Monitoring Mission long-range UAV was unable to take off from a Ukrainian airbase in Stepanivka due to GPS signal interference (OSCE, 2021). In addition, Russian jamming of GPS signals in Ukraine has been detected by US forces in the region (Hitchens, 2022).

On the February 24, 2022, a cyberattack against a commercial satellite network belonging to the US company Viasat, not only had an impact on Ukrainian military actors but also damaged the terminals of civilian customers across Europe and affected thousands of wind turbines in Germany (ESPI, 2022). Tens of thousands of satellite modems had their internet service knocked out after being flooded with traffic along with destructive commands to overwrite key data. This highlights the wider impact that cyberattacks can have on the satellite industry.

Jamming of Satellite Communications

"Traditional" satcom jamming is another counterspace area where Russia have invested significant effort. The R-330Zh "Zhitel" mobile jammer is reported to able to jam commercial Inmarsat and Iridium receivers within a tactical local area. The TsNII research institute has declared that Tirada-2S was under development and will be used to conduct uplink jamming of comsats (Hendrickx, 2020a). It is likely Tirada-2S is currently in service. Another system under development is Bylina-MM, which is designed to suppress the on-board transponders on comsats such as Milstar, Skynet and Italsat (Weeden & Samson, 2021). Another key project in Russia's EW programme is TOBOL designated 14Ts227, with a project infrastructure code of 8282. Indications about the goals of Tobol suggest that the site would have an array of ground-based antennas that would pick up and jam what are called unauthorized signals sent to satellites or relayed via satellites to the ground. Vatutin, who heads a department within Russian Space Systems and is identified as Tobol's chief designer, has co-authored several papers and patents related to the protection of satellites from electronic attack. One such patent describes an array of ground-based antennas that would be used to pick up and jam unauthorized signals sent to satellites relayed via satellites to the ground (Hendrickx, 2020a). In another scenario, unauthorized signals downlinked from a satellite to the ground would be identified by monitoring stations, following which the tropospheric stations would transmit jamming signals that would reach receivers after being reflected off the troposphere and cancel out the effects of the unauthorized signals (Hendrickx, 2020b). Another paper co-authored by Vatutin discussed the possibility of using EW techniques to prevent both optical and radar reconnaissance satellites from sending images to data relay satellites as they fly over. This reflects growing interest in the use of EW systems to counter foreign reconnaissance assets. Finally, the Krashuka-4 mobile EW system, designed to counter airborne early warning and control and other airborne radar, has an effective range of 300 kilometres. Due to its range and power, it is also effective against LEO synthetic aperture radar (SAR) imaging satellites (Weeden & Samson, 2021).

Directed Energy Weapons

Russia has a long history of research in high-energy laser physics. Directed Energy Weapons (DEW) is a ranged weapon that damages a target with highly focused energy without a solid projectile. Russia revived its old Soviet Airborne Laser system in 2012, called Sokol-Echelon. The Russian ABL was designed to counter space-based reconnaissance assets in the infrared part of the spectrum, dazzling rather than destroying (Hendrickx, 2020a). The laser type selected was a carbon monoxide laser. In mid-2018 a court document declared that the MOD had decided to cancel the project in late 2017, however, contracts signed as part of the project continue to appear on the Russia's government procurement website afterwards.

Russia is upgrading its Krona optical surveillance system in the North Caucasus with laser dazzling capabilities. The Krona complex historically included ground-based radars and optical telescopes for tracking, identifying and characterizing space objects. Under a project code-named Kalina for the Ministry of Defence, its goal was the creation of a channel for the suppression of electro-optical systems of satellites using solid-state lasers (Hendrickx, 2020a). Russia is also planning to develop a laser with a range of 40,000 kilometres to attack early warning satellites in geosynchronous orbit (Liu et al., 2020).

Russia Pursues Nuclear Weapons in Space

In February 2024, it was publicly released by US officials that Russia was pursuing the development of a space-based ASAT weapon equipped with a nuclear device. Clarification came from the White House National Security Council spokesman John Kirby that it was "not an active capability that's been deployed" (Breaking Defense, 2024). Further details were not elaborated, however, he did confirm it was "related to an anti-satellite weapon that Russia is developing." This action would be in clear breach of the 1967 Outer Space Treaty which prohibits the deployment of weapons of mass destruction in space.

The Russia satellite referred to is Kosmos-2553, which was launched on February 5, 2022. The Russian Ministry of Defence referred to it as a "technological satellite equipped with newly developed on-board instruments and systems in order test them in conditions of radiation and heavy particle charged particles" (FFF, 2022). This is likely a cover story for Kosmos-2553, a likely nuclear mission. It was launched into a 1,987 by 1,995 km orbit with an inclination of 67 degrees. The detonation of a nuclear weapon in LEO, and the subsequent nuclear electromagnetic pulse (EMP) effects could render that orbital regime unusable for up to a year if not longer. US intelligence agencies have assessed that Kosmos-2553 was a practice test run for putting a nuclear weapon into orbit (Hendrickx, 2024).

The Russian development of a nuclear weapon in orbit in space is seen as the only effective way of countering mega-constellation such as Starlink or indeed the US Space Development Agency's Tracking Layer concept which sees the development of hundreds of satellites for the National Defence Space Architecture. What is clear is any detonation in LEO of a Russian nuclear weapon in space would have a significant indiscriminate effect on satellites in LEO for a significant period of time.

Chinese Counterspace Program

China has a long history of developing space weapons. In 2007, China demonstrated its capability to kinetically intercept satellites in LEO from the ground (SW Foundation Factsheet, 2012). In 2022, China demonstrated a novel, game changing, capability to hide in the "graveyard" beyond geostationary orbit (GEO) and re-emerge to grapple a satellite in GEO (Space News, 2022). Additionally, it has the ability to use ground-based lasers to dazzle satellites in LEO. China has the ability to conduct radiofrequency (RF) jamming from mobile platforms against communication satellites in LEO.

Despite Chinese statements on space warfare claiming to adhere to the peaceful uses of outer space, China has designated space as a military domain since 2015 (Pollpeter, 2016). The goal of space warfare and operations is to achieve space superiority. Chinese writings as early as 2012 have declared the need for space dominance. Examples include: "Successful efforts at establishing space dominance therefore must also take into account the sustainment of this entire structure of terrestrial and space systems and associated data and communications links, while striving to degrade or destroy an opponent's" (Cheng, 2012), and "Therefore, attacks against strategic space targets require the direction of the highest-level political authorities" (Cheng, 2012).

Co-Orbital Assets

In the summer of 2010 SHI JIAN-12 (SJ-12) conducted a number of close approaches with the SJ-06F satellite in LEO (600 to 570 kilometres.) These occurred between June and August 2010. In the closest approach, the two satellites were less than 300 meters apart (Weeden & Samson, 2021). SJ-12 was launched into a higher inclination, then manoeuvred in August 2010 to be essentially the same inclination as SJ-06F.

Another proximity operation occurred in LEO in 2013. Three payloads were placed in orbit at 670 kilometres from the same launch on July 19, 2013: SHIYAN-7, CHUANGXIN-3 and SHIJIAN-15. SY-7 was known to the Chinese program as TANSUO-4 and was likely fitted with a robotic arm, which interacted with a separating subsatellite, known as TANSUO-3 (CX-3), and was designed to provide optical surveillance in GEO and LEO (Weeden & Samson, 2021). SJ-15 known as TANSUO-5 was designed to manoeuvre and conduct proximity operations.

On August 9, 2013, TANSUO-5 manoeuvred close to TS-3 and TS-4 passing close to TS-3 at a distance of a few kilometres. TS-5 on August 16 altered its altitude by more than 100 kilometres to conduct a close approach within a few kilometres of SHI JIAN (SJ-7), which was launched in 2005.

On October 18, TS-4 raised its orbit by several hundred meters and released a small object, which was designated as debris. TS-4 and the "debris" orbited in close proximity for several days, ranging between a few kilometres and several hundred meters and it was reported that the two objects were joined together, with TS-4 clasping the smaller "debris" object. These two objects conducted small manoeuvres during 2014–15 with the distance never exceeding a few kilometres (Weeden & Samson, 2021).

In April 2014, SJ-15 (TS-5) began to conduct manoeuvres around TS-3. Between May 12 and 14, TS-5 manoeuvred by lowering to match orbital planes with SJ-7, and on a trajectory that brought it above and then behind SJ-7 at a range of around 150 kilometres with a vertical separation of a few kilometres. During May, TS-5 reduced the distance to SJ-7 to within a kilometre. In 2015–16, TS-5 (SJ-15) occasionally made changes to its orbit however with no apparent reason, except for perhaps demonstrating technologies to perform proximity operations.

In 2016, the AOLONG-1 (AL-1) small satellite, known as the Advanced Debris Removal Vehicle (ADRV), demonstrated using a robotic arm to capture a small piece of debris for removal from orbit. The AL-1 did not conduct any rendezvous operations with any objects and it did not appear to change its orbit during its two months on orbit (Weeden & Samson, 2021). The satellite used a robotic arm to grapple virtual targets (GSS).

On November 3, 2016, China launched the SHIJIAN-17 (SJ-17) into GEO. SJ-17 was reportedly designed to test advanced technologies, however it was also fitted with an onboard optical surveillance sensor (Weeden & Samson, 2021) and a reported signals intelligence mission (Chen, 2017). Unusually, the YZ-2 upper stage remained in obit with a perigee near GEO, and will dip down very close to and rotate around the active GEO belt for decades to come. On November 11, 2016, SJ-17 manoeuvred to 162.9 degrees east in GEO, relatively close to CHINASAT-5A, a Chinese communications satellite, coming as close as few kilometres by November 30. At the end of 2016, SJ-17 drifted away and began a rapid drift east of two degrees per day, and four degrees west per day on February 9, 2017, until on March 20 it lowered its orbit and moved to rendezvous and proximity operations (RPO) with CHINASAT-20, a military communications satellite (Weeden & Samson, 2021).

On July 29, 2018, SJ-17 conducted a RPO with CHINASAT-1C, a communications satellite that had an unexplained anomaly and had begun drifting westward at 0.5 degrees per day. On January 23, 2018, SJ-12 raised its inclination from 0.43 to roughly four degrees, and manoeuvred to a drift of four degrees per day, before reversing back to zero between July 20 and 22, 2018. The large change in inclination suggests the SJ-17 has significant dV capability as plane change manoeuvres are amongst the most energy intensive. SJ-17 came within 1.5 kilometres of CHINASAT-1C and highly likely conducted inspection for ten days to assess the source of the anomaly and monitor the recovery attempt of CHINSAT-1C (Weeden & Samson, 2021). In 2020, SJ-17 made RPO with CHINASAT-6B in January, and with SJ-20 in October.

On December 23, 2018, China launched another mission to GEO, the Tongxin Jishu Shiyan (TJS)-3. Two objects were catalogued from the launch, TJS-3 and a second object, 43917. What was unusual about 43917 is that in it did a series of manoeuvers to place it into GEO at 59.07 degrees east near TJS-3. It appears that 43917 is subsatellite and maintaining a close distance, about 100–200 kilometres, from TJS-3. In April 2019, TJS-3 left object 43917 and moved to another location, suggesting initial checkout took place near object 43917 (Weeden & Samson, 2021). This is a significant program of testing in LEO and GEO undertaken by the Chinese government that goes beyond what could be deemed necessary for satellite inspection, space debris removal and space situational awareness. It is likely these programs could be part of wider counterspace capabilities. As a Defense Intelligence Agency (DIA) report concluded "China is developing sophisticated on-orbit capabilities, such as satellite inspection and repair, at least some of which could also function as a weapon" (DIA, 2019). The ambiguity of RPO for in-orbit servicing or counterspace weaponry lies in the fact that the onboard tracking and guidance could be used to collide with another satellite to damage or destroy it. A caveat is that, to date, the Chinese RPO satellites would need higher relative velocities and longer closing distances (Weeden & Samson, 2021). While debris removal and in-orbit servicing could be characterized as benign activities, such assets are now directed by the PLA Strategic Support Forces (Goswami, 2019). This indicates that activities which were once under civilian purview could be switched when required as the program is directed and led by the PLA. A further potential offensive use of RPO would be to install a RF jammer onboard the chaser satellite, increasing its ability to interfere with the satellite's communications. Chinese academic papers recognized that reducing the distance with a small satellite platform would decrease the power requirements exponentially, identifying susceptible US assets such as the Advanced Extremely High Frequency satellites (Chen, 2017). This, coupled with the Chinese doctrine that China can defeat the United States "network centric warfare" with "energy-centric warfare," indicates that China has a significant interest in developing high-frequency and directed energy weapons in space.

On October 24, 2021, China launched Shijian-21 (SJ-21) with China's state-run Xinhua news agency reporting its mission as "mainly used to test and verify space debris mitigation technologies" (NASA NSSDC). On November 1, a new object was catalogued (NASA, 2025b) alongside SJ-21, which is potentially its AKM (Apogee Kick Motor) as catalogued by the US Space Force 18th Space Control Squadron. There has been speculation that this object alongside SJ-21 could be a sub-satellite released after arriving in GEO (Burke, 2021). This speculation is based on the fact that both objects appear to remain five kilometres apart and deliberately synchronized after completing a re-rendezvous on November 15 (Planet 4589, 2021). If the AKM was ejected, a steadily increasing separation would be expected. The nature of the proximity operations of the AKM and SJ-21 is not known and it could be part of China's counterspace testing. On January 22, 2022, SJ-21 was observed to execute a large manoeuvre to bring itself closely next to a dead Beidou navigation satellite. SJ-21 pulled the dead satellite out of its geosynchronous orbit and placed it a few hundred miles into a graveyard orbit. SJ-21 appeared to be acting as a space tug. On January 26, SJ-21 released the Beidou and manoeuvred back near GEO. The capabilities the SJ-21 demonstrated could be used in a counterspace role to move other satellites in geosynchronous orbit. The technological demonstration of the SJ-21, while perhaps having legitimate purposes to remove defunct satellites with depleted fuel reserves to the graveyard orbit, could also serve a counterspace role.

Direct Ascent Assets

The Chinese direct ascent program's first known tests were in 2005 and 2006, using the SC-19, also referred to as Dong Neng-1 (DN-1), and is likely a variant based on the DF-21 mobile series of ballistic missiles. On January 11, 2007, the DN-1 launched from Xichang and successfully destroyed a defunct Chinese Feng Yun-1C weather satellite at an altitude of 865 kilometres (Weeden & Samson, 2021). This ASAT test created a large amount of debris (SW Foundation) and generated a significant amount of international condemnation.

On May 13, 2013, a likely test of a DA-ASAT that could reach higher orbits took place from Xichang (Weeden & Samson, 2021). This test was reported by the Chinese Academy of Sciences as a high-altitude scientific research mission. The Notice to Airmen (NOTAM) released by China that provided advance warning of the flight path covered a ground track which lined up with a GEO launch trajectory. Also, technical analysis conducted by the Union of Concerned Scientists, based on it re-entering above the Indian Ocean, indicated that the test had an apogee of 30,000 kilometres with a flight time of 6.7 hours (Weeden, 2014). This is consistent with US military official statements at the time of "nearly to GEO". This new ASAT test variant was labelled DN-2, with an estimated operational timeframe of 2020–25. A recent National Air and Space Intelligence Center report stated that "China has military units that have begun training with anti-satellite missiles" (NASIC, 2018). More recently, the Director of National Intelligence reported in 2019 that: "The People's Liberation Army (PLA) has an operational ground-based ASAT missile intended to target low-Earth-orbit satellites, and China probably intends to pursue additional ASAT weapons capable of destroying satellites up to geosynchronous Earth orbit" (Coats, 2019). It is likely that DN-1 is operational as this is intended to target satellites in LEO, however, DN-2 is likely in development perhaps by 2025.

Electronic Warfare as Counterspace

China has significant capabilities in global navigation satellite system (GNSS) jamming capabilities and has developed both fixed and mobile systems. The known systems are downlink jammers, which can affect GNSS receivers in a local area (Weeden & Samson, 2021). There is no open source literature that indicates that China can target the uplink jamming of GNSS satellites. A US vessel travelling in Shanghai in 2019, the Manukai, reported that its GPS systems had been jammed at the berth, as both of the ship's GPS units had lost their signals, and its AIS transponder had failed. A last-ditch emergency system, like AIS, also depended on GPS could not get a fix (Harris, 2019). Indeed, there was evidence that the GPS systems had been spoofed as its true position and speed had been replaced by false coordinates broadcast from the ground.

According to open source data in April 2018, China installed equipment capable of jamming communications and radar systems on two of its fortified outposts on the Spratly Islands in the South China Sea (Gordon & Page, 2018).

The PLA during exercises routinely incorporates jamming and anti-jamming techniques against multiple communication, radar systems and GPS satellite systems in exercises. A Defense Intelligence Agency report assessed that China is developing jammers to target SATCOM over a range of frequency bands including military protected extremely high frequency communications (DIA, 2019).

Given the importance of EW in Chinese military doctrine, it is likely that China is developing a ground-based synthetic aperture radar (SAR) jamming capability. The former GSD Third Department (now SSF) oversaw a division leader-grade unit, headquartered in Shanghai, responsible for the intercept of Satcom and SAR transmissions (Stokes, 2019).

Directed Energy Weapons

China is actively pursuing the development of directed energy weapon (DEW) for counterspace use. There is a significant amount of evidence of research and development, and testing but limited details on operational status of any deployed capabilities (Weeden & Samson, 2021). The use of lasers as a weapon is characterized in three effects: dazzling a satellite's imaging sensor, and/or damaging a satellite's imaging sensor and/or damaging to the satellite bus or subsystems. The effect of dazzling is temporary, and is considered a countermeasure rather than a weapon. Relative low power levels are required to dazzle. A 10-watt laser could be sufficient to create a dazzling effect and obscure an area from being imaged (Weeden & Samson, 2021). The threshold between dazzling and damage is almost impossible to predict, as it would depend on knowledge of a target satellite's internal design and protective mechanisms. For use as a weapon to cause significant damage to the sensor, a power level in the kilowatt range would be required. A very-high-power laser would be required to cause damage to the satellite bus. The damage would be due to the heating effects of the energy causing the essential components such as the thermal regulation system, the batteries or attitude control system. It was reported in 2006 that China used a ground-based laser to dazzle or "blind" a US optical surveillance satellite on at least one occasion (Space News, 2006). China has at least five sites that support China's DEW work:

- Two sites at the Centre for Atmospheric Optics, Anhui Institute for Optics and Fine Mechanics in Hefei
- Chinese Academy of Engineering Physics in Mianyang, Sichuan Province
- Korla Missile Test Facility in Xinjiang Province

Similarly, commercial imagery of Xijiang Province has shown a similar layout to the other facilities, indicating DEW research is being conducted. The facility has four main buildings with sliding roofs, with three of the sheds connected with two vacuum spheres. The shape and size indicate that possibly chemical lasers are being used (Bhat, 2019). It is postulated that the equipment under the smaller sliding shed is used for tracking, while the other three are used individually or in conjunction with each other. Both the Hefei and Mianyang facilities have similarly large rectangular buildings with retractable roofs and suggest facilities where DEW aimed at satellites could have been developed (Weeden & Samson, 2021).

Chinese ground-based laser testing was possibly confirmed by a Chinese article from scientists working at the Changchun Institute of Optics, Fine Mechanics and Physics:

> In 2005, we have successfully conducted a satellite blinding experiment using a 50-100 KW capacity mounted laser gun in Xinjiang province. The target was a low orbit satellite with a tilt distance of 600 km. The diameter of the telescope firing the laser beam is 0.6 m wide. The accuracy of ATP (acquisition, tracking and pointing) is less than 5 [microradians].
>
> (Fisher, 2017)

The site of the ground-based laser was likely located at Korla, Xinjiang province. It is likely that China has developed more powerful ground-based lasers since this test (Fisher, 2017). A recent DIA assessment assessed that China possibly already has a "limited capability to employ laser systems against satellite sensors" (DIA, 2019). It postulated that China may field a higher power system by the mid-to-late 2020s capable of threatening the structure of non-optical satellites (DIA, 2019).

There is limited reporting from the Changchun Institute of Optics, Fine Mechanics and Physics scientists article examining the cancelled US ABL in 2011 that China has an airborne laser program for ASAT use. In the Chinese Communist Party celebrations in 2009 a museum displayed an image of a four-engine aircraft using a laser to attack a satellite (Fisher, 2017). This could indicate that China is continuing to conduct research into the feasibility of using an ABL in an ASAT capability. The same Changchun authors argue for the feasibility of a Chinese space-based laser weapon. The authors concluded that

> In future wars, the development of ASAT weapons is very important. Among those weapons, laser attack system enjoys significant advantages of fast response speed, robust counter-interference performance and a high target destruction rate, especially for a space-based ASAT system. So the space-based laser weapon system will be one of the major ASAT development projects.
>
> (Fisher, 2017)

While the ability of China to develop such a system in the stated timeframe is in question, the desire to develop such a system exists with the technical space community.

Table 5.1 China and Russia's Space-Based ASAT Testing Programmes

Launch Date	*Country*	*Satellite*	*System*
December 25, 2013	Russia	Cosmos 2488, 2489, 2490, 2491	Nivelir ASAT development
May 2023	Russia	Cosmos 2496, 2497, 2498, 2499	Nivelir ASAT development
June 23, 2019	Russia	Cosmos 2519, 2520, 2521, 2523	Nivelir ASAT development
July 10, 2019	Russia	Cosmos 2542, 2543	Nivelir ASAT development
February 5, 2022	Russia	Cosmos 2533, 1558	Nivelir ASAT development
May 16, 2024	Russia	Cosmos 2576	Nivelir ASAT development
2013	China	Tansuo 3, 4, 5	Robotic Arm
November 3, 2016	China	Shijian-17	Optical and Sigint Proximity Operations Testing possibly EW
December 23, 2018	China	TJS-3	Co-orbital ASAT Development
October 24, 2021	China	SJ-21	ASAT Development for Grappling and Proximity Operations

Conclusion

Both China and Russia are rapidly advancing counterspace capabilities. The pace of development greatly surpasses the international community's ability to monitor and counter-react. A much faster, more agile, risk-taken approach needs to be taken to allow innovative solutions to be created to tackle the novel space threats. All states need to make the protection of space assets their number one priority but all efforts need to be co-ordinated and in coherence. A potential solution needed to address this space threat is the deployment of a multi-orbit system of systems that can detect non-ballistic missiles from launch to their designated target areas (Stone, 2022). A multi-layered, space-based architecture in LEO, MEO, GEO and polar orbits is recommended to not only detect missiles at all altitudes but also to provide fire control information in near real time. Resilience could be further enhanced by deploying satellites that are capable of enhanced manoeuvre to avoid ASATs; this capability tends towards smaller detection satellites in LEO. Additionally, deploying decoys in LEO, MEO and GEO would complicate an adversary's attacks.

References

Antrobus, A. 2020. "Wake-up Call for Space Threats." *Royal Aeronautical Society*, www.aerosociety.com/news/wake-up-call-for-space-threats/.

Bhat, V. 2019. "These Futuristic Chinese Space Denial Weapons Can Disable or Destroy Opposing Satellites." *The Print*, March 23. https://theprint.in/defence/these-futuristic-chinese-space-denial-weapons-can-disable-or-destroy-opposing-satellites/210212/.

Breaking Defence. 2024. "Russia's Nuclear Weapon in Space: Mike Turner Threat White House." *Breaking Defence*, February. https://breakingdefense.com/2024/02/russia-nuclear-weapon-space-mike-turner-threat-white-house/.

Burke, K. 2021. "China's SJ-21 Framed as Demonstrating Growing On-Orbit Servicing, Assembly, and Manufacturing (OSAM) Capabilities." *China Aerospace Studies Institute*, December 6, www.airuniversity.af.edu/Portals/10/CASI/documents/Research/Space/2021-12-09%20SJ-21%20and%20China's%20OSAM%20Capabilities.pdf.

Chen, D. 2017. "Hearing on 'China's Advanced Weapons,' Panel on China's Directed Energy and Electromagnetic Weapons Programmes." February 23. www.uscc.gov/hearings/hearing-chinas-advanced-weapons.

Cheng, D. 2012. "China's Role in Space." *Strategic Studies Quarterly*, 6 (1): 55–77. www.jstor.org/stable/26270790?seq=1.

Coats, D. 2019. "Director of National Intelligence, Statement for the Record, Worldwide Threat Assessment of the US Intelligence Community." Senate Intelligence Committee, January 29.

DIA. 2019. *Challenges to Security in Space*, January. https://aerospace.csis.org/wp-content/uploads/2019/03/20190101_ChallengestoSecurityinSpace_DIA.pdf.

Episkopos, M. 2021. "GPS Jamming: Can NATO Defeat This Russian Weapon in the Arctic?" *National Interest*, March 3. https://nationalinterest.org/blog/reboot/gps-jamming-can-nato-defeat-russian-weapon-arctic-179143.

ESPI. 2022. "The War in Ukraine and the European Space Sector." *Executive Brief no.57*. May 5. www.espi.or.at/briefs/the-war-in-ukraine-and-the-european-space-sector/.

Final Frontier Flash, ISR University, 13 February 2022. https://isruniversity.com/wp-content/uploads/2025/02/2022-02-13-Final-Frontier-Flash.pdf

Fisher, R. D. 2017. "China's Progress with Directed Energy Weapons Testimony to Congress." February 23. www.uscc.gov/sites/default/files/Fisher_Combined.pdf.

Gordon, M. R., and J. Page. 2018. "China Installed Military Jamming Equipment on Spratly Islands, US Says." *Wall Street Journal*, April 9. www.wsj.com/articles/china-installed-military-jamming-equipment-on-spratly-islands-u-s-says-1523266320.

Goswami, N. 2019. "Before the U.S.-China Economic and Security Review Commission Hearing on 'China in Space: A Strategic Competition?'" April 25. www.uscc.gov/sites/default/files/2019-10/April%2025%202019%20Hearing%20Transcript.pdf.

Harding, T. 2023 "'Satellite Bodyguards' Prepared for Space Protection." *The National*, January 25. www.thenationalnews.com/world/uk-news/2023/01/25/satellite-bodyguards-prepared-for-space-protection/.

Harris, M. 2019. "Ghost Ships, Crop Circles, and Soft Gold: A GPS Mystery in Shanghai." *MIT Technology Review*, November 15. www.technologyreview.com/2019/11/15/131940/ghost-ships-crop-circles-and-soft-gold-a-gps-mystery-in-shanghai/.

Hendrickx, B. 2020a. "Peresvet: A Russian Mobile Laser System to Dazzle Enemy Satellites." *The Space Review*, June 15. www.thespacereview.com/article/3967/1.

Hendrickx, B. 2020b. "Russia Gears Up for Electronic Warfare in Space (Part 1)." *The Space Review*, October 26. www.thespacereview.com/article/4056/1.

Hendrickx, B. 2024. "Russian Research on Space Nukes and Alternative Counterspace Weapons Part 1." *The Space Review*, May 13. www.thespacereview.com/article/4793/1.

Hitchens, T. 2022. "Local Russian GPS Jamming in Ukraine Hasn't Affected US Support Ops, So Far." *Breaking Defence*, March 1. https://breakingdefense.com/2022/03/local-russian-gps-jamming-in-ukraine-hasnt-affected-us-support-ops-so-far/.

Liu, Zhenhua, Chuanwen Lin, and Gang Chen. 2020. "Space Attack Technology Overview." *Journal of Physics; Conference Series,* 1544. https://iopscience.iop.org/article/10.1088/1742-6596/1544/1/012178

Mowthorpe, M. 2022. "The Russian Space Threat and a Defense Against It with Guardian Satellites." *The Space Review*. June 13. www.thespacereview.com/article/4401/1.

NASA. 2025a. "Cosmos 2491." March 20. https://nssdc.gsfc.nasa.gov/nmc/spacecraft/display.action?id=2013-076E.

NASA. 2025b. "2021-094A." March 20. https://nssdc.gsfc.nasa.gov/nmc/spacecraft/display.action?id=2021-094A.

NASIC. 2018. *Competing in Space*, December. www.nasic.af.mil/About-Us/Fact-Sheets/Article/1738710/competing-in-space/.

NATO. 2019. "Foreign Ministers Take Decisions to Adapt NATO, Recognize Space as an Operational Domain." November 20. www.nato.int/cps/en/natohq/news_171028.htm.

OSCE. 2021. "Spot Report 6/2021: SMM Long-Range UAV Unable to Take Off Due to Dual GPS Signal Interference." April 7. www.osce.org/special-monitoring-mission-to-ukraine/483008.

Planet 4589. 2021. "Jonathan's Space Report." November 28. https://planet4589.org/space/jsr/back/news.800.txt.

Pollpeter, K. 2016. "Space, the New Domain: Space Operations and Chinese Military Reforms." *Journal of Strategic Studies,* 39 (5-6): 709–727.

Reuters. 2022. "Russia Warns West: We Can Target Your Commercial Satellites." October 27. www.reuters.com/world/russia-says-wests-commercial-satellites-could-be-targets-2022-10-27/.

RussianSpaceWeb. 2024. "Soyuz-2-1v Launches a Secret Satellite." March 12, www.russianspaceweb.com/napryazhenie.html.

Secure World Foundation Factsheet 2012. https://swfound.org/media/9550/chinese_asat_fact_sheet_updated_2012.pdf

Space News. 2006. "NRO Confirms Chinese Laser Test Illuminated US Spacecraft." October 3. https://spacenews.com/nro-confirms-chinese-laser-test-illuminated-us-spacecraft/.

Space News. 2022. "China's Shijian-21 Spacecraft Docked with and Towed a Dead Satellite." January 27. https://spacenews.com/chinas-shijian-21-spacecraft-docked-with-and-towed-a-dead-satellite/.

Stokes, M. 2019. "US Hearing on 'China in Space: A Strategic Competition.'" April 25. www.uscc.gov/sites/default/files/2019-10/April%2025%202019%20Hearing%20Transcript.pdf.

Stone, C. 2022. "Orbital Vigilance: The Need for Enhanced Space-Based Missile Warning and Tracking." *Mitchell Institute Policy Paper*, June. www.mitchellaerospacepower.org/orbital-vigilance-the-need-for-enhanced-space-based-missile-warning-and-tracking/.

The Guardian. 2020. "Britain and US Accuse Russia of Launching 'Weapon' in Space." July 23. www.theguardian.com/world/2020/jul/23/britain-us-accuse-russia-launching-weapon-space-satellite-threat.

The War Zone. 2019. "Russia Jammed Phones and GPS in Northern Europe During Massive Military Drills." June 30. www.twz.com/15194/russia-jammed-phones-and-gps-in-northern-europe-during-massive-military-drills.

The Washington Post. 2021. "Russia Proved It Can Shoot Down a Satellite. Does This Make Space Less Secure?" November 24. www.washingtonpost.com/politics/2021/11/23/russia-proved-it-can-shoot-down-satellite-does-this-make-space-less-secure/.

Trichas, M. 2024. "Opportunities and Threats from Commercial Satellite Mega-Constellations." *NATO OPEN*, https://issuu.com/spp_plp/docs/open_publications_opportunities_and_threats_from_c.

UK. Gov. 2022. "Russia Behind Cyber-Attack with Europe-Wide Impact an Hour Before Ukraine Invasion." May 10. www.gov.uk/government/news/russia-behind-cyber-attack-with-europe-wide-impact-an-hour-before-ukraine-invasion.

Weeden, B. 2014. "Through a Glass, Darkly: Chinese, American, and Russian Anti-Satellite Testing in Space." *Secure World Foundation*, March 17. https://swfound.org/media/167224/through_a_glass_darkly_march2014.pdf.

Weeden, B. 2015. "Dancing in the Dark Redux: Recent Rendezvous and Proximity Operations in Space." *The Space Review*, October 5. www.thespacereview.com/article/2839/3.

Weeden, B., and V. Samson. 2021. "Global Counterspace Capabilities." *Secure World Foundation*, April, https://swfound.org/media/207161/swf_global_counterspace_capabilities_es_2021_en.pdf.

6 The Role of Hypersonics in Modern Warfare

Tracey German

Introduction

In the rapidly evolving landscape of modern warfare, the emergence of hypersonic technology has ushered in a new era of strategic capabilities and military supremacy. Hypersonic weapons systems, with their ability to strike targets with unparalleled speed and precision, have not only raised questions over deterrence but have also challenged traditional notions of defence and offence. This chapter explores the role that hypersonic weapons could play in reshaping the dynamics of conflict. Hypersonic weapons are perceived to sit alongside other modern technological advances such as AI, cyber and UAVs in terms of the potential for changing the face of the battlefield. The director of the US Defence Intelligence Agency has argued that developments in 'hypersonics propulsion will revolutionise warfare by providing the ability to strike targets more quickly, at greater distance, and with greater firepower' (Defense Intelligence Agency 2018). But what is the reality? Will hypersonic weapons revolutionise the conduct of war? And what are the strategic implications of hypersonic weapons and their proliferation? Hypersonic weapons were used on the battlefield for the first time in March 2022, when Russia struck targets in Ukraine with its Kh-47M2 Kinzhal hypersonic missile. However, even before this first use they had entered the popular imagination, with Sam Fender singing about hypersonic missiles on his debut album of the same name in 2019. The advent of hypersonic weapons technology has generated a lot of noise, but there are doubts over whether the early use of these missiles on the battlefield has lived up to the promise. Russia has got through its limited inventory relatively quickly, generating little substantive change in the military situation. This chapter sets out the key characteristics of hypersonic missiles, the challenges presented by such weapons and explores the Russian example to demonstrate why states decide to pursue hypersonic technology, as well as the potential pitfalls.

Williams (2020) notes that the arrival of aerial hypersonic weapons shares some similarities to the introduction of other long-range strike systems, such as the intercontinental ballistic missile (ICBM) and the strategic bomber. He cites Herman Kahn, who warned in the 1960s that ICBMs and other nuclear weapons could make wars much easier to start, as their constant readiness would obviate

DOI: 10.4324/9781003520160-6

the need to mobilise for a major military effort.[1] Although these technological advances changed the character of conflict, they were not as disruptive as initially feared. Many analysts view the emergence of hypersonic weapons in a similar light, concerned that they may disrupt strategic stability and the balance of military power (Wilkening 2019). By contrast, Reny (2020) has reasoned that nuclear-armed hypersonic weapons will provide an overall stabilising effect in the global arena, but will further destabilise regional competition and conflict. A 2019 UN report posits that although the military utility of hypersonic weapons systems remains unclear, they may offer new capabilities that have strategic ramifications, even if the weapons themselves are not strategic in nature (United Nations Office for Disarmament Affairs 2019, viii). It is important to note that, currently, hypersonic missiles are not yet nuclear armed; they only offer conventional capabilities.

The development of a hypersonic weapons capability needs to be viewed as part of a state's pursuit of strategic advantage over a competitor or adversary. Stone (2018) has also compared the hypersonic arms race as a 'race to the moon sort of thing', involving national pride. Thus, the pursuit of hypersonic capability is linked to potential arms racing and the notion of a security dilemma. While hypersonic weapons offer a number of advantages, notably speed and precision, they are also very expensive and therefore out of reach for many states. As will be discussed below, the US, Russia and China have been developing hypersonic weapons, driven to a large extent by fear of losing strategic advantage to their competitors. This was emphasised by the US Department of Defense, which has noted the need to 'sustain military advantage over China', by making major investments in a range of areas, including integrated air and missiles defences, and hypersonics (Office of the Under Secretary of Defense 2023, ii).

Defining Hypersonic

The term 'hypersonic' refers to travelling at five times the speed of sound, Mach 5, around 3,700 mph. There are a number of challenges associated with hypersonic flight, including the fact that aerodynamic forces at high speeds are severe. Furthermore, hypersonic flight produces intense heat, complicating the production of viable weapons systems. While hypersonic speed is a defining characteristic that differentiates hypersonic missiles from others, hypersonic weapons are not defined solely by the speed at which they travel; Oelrich (2020) points out that modern ICBMs already travel above Mach 5 and thus hypersonic weapons have been in use for decades. However, ICBMs fly on a predictable trajectory as they are unpowered. The difference between hypersonic missiles and ballistic missiles are the former's key characteristics of speed, range and manoeuvrability. This manoeuvrability means that the missile's final destination remains ambiguous until the target is reached, making them difficult to defend against as they need to be continuously tracked, while the speed of their flight means that response timelines are compressed. Hypersonic weapons also fly in a different part of the atmosphere to existing missiles – higher than subsonic missiles, but lower than ICBMs – further complicating efforts to track their trajectory. The advantages offered by hypersonic weapons were set out in a report from the US Department of Defense, which stated that '[h]ypersonic systems expand our ability to hold distant targets at risk, dramatically shorten the timeline to strike a target, and their manoeuvrability increases survivability and unpredictability', noting that strategic competitors are rapidly developing advanced hypersonic missiles (Office of the Under Secretary of Defense 2023, 4–24).

Hypersonic Weapons Systems

There are two main types of hypersonic weapons systems: hypersonic cruise missiles (HCMs) and hypersonic glide vehicles (HGVs). HCMs maintain a constant hypersonic speed (and usually altitude) and are powered over the entire course of their flight, initially boosted by a rocket and then using an air-breathing engine known as a scramjet to maintain their speed. Consequently, HCMs require smaller launch rockets than HGVs and therefore cost less. The US successfully tested a scramjet hypersonic missile, the Hypersonic Air-breathing Weapon Concept (HAWC), in 2022 (Liebermann 2022). HGVs are typically launched on top of ballistic missiles (often referred to as a boost-glide system) and then glide back through the atmosphere to their target at hypersonic speeds (SIPRI 2022). Examples of HGVs include China's Dongfeng-17 (DF-17), Russia's Avangard and the U.S. Navy's Conventional Prompt Strike system.

The discussion about hypersonic weapons is often complicated by a blurring of the lines between conventional and nuclear capabilities. Although they are not (currently) nuclear-armed, hypersonic missiles bolster conventional capabilities and could enhance nuclear deterrence and associated strategic stability. There is also debate about the utility of hypersonic weapons and the type of missions that hypersonic weapons could be deployed on: what missions require the capability to project weapons globally? Acton (2013, 7) suggests there could be four possible missions for hypersonic missiles, which require a rapid, unexpected response: denying an adversary the ability to utilise its nuclear arsenal, disabling satellite capabilities, impeding A2AD and striking high-value targets such as an aircraft carrier, the destruction of which is likely to have a significant impact on the outcome of a major war.

Notwithstanding the hype surrounding their development, hypersonic weapons present a range of challenges, from the difficulties associated with their development to the question of how to defend against them. The production of hypersonic missiles presents a range of technical challenges, necessitating complex technological solutions, and are therefore very expensive to develop. Consequently, they are not currently produced in large numbers. Unpredictable trajectories lead to target ambiguity, making them very difficult to defend against and prompting concern that this ambiguity could lead to strategic instability. Their speed compresses timescales for decision-makers in states that are being targeted, making it much more difficult to detect, assess, decide and take counter actions.

Impact on Deterrence and Strategic Stability

The speed and manoeuvrability of hypersonic missiles means that they are able to penetrate existing ballistic missile defence (BMD) systems, as well as most air defences, undermining a state's ability to defend itself. Without accurate and timely information about the precise path a missile is taking, it is almost impossible to defend against an attack. The development of hypersonic weapons technology is likely to predominantly impact states that already have effective BMDs, potentially shifting the existing balance of military power and impacting strategic stability. Existing ballistic and cruise missiles are considered to be a cost-effective weapon and symbol of national power, particularly if nuclear-armed (Defence Intelligence Ballistic Missile Analysis Committee 2020, 2). The advent of nuclear weapons, in particular the development of nuclear-armed Intercontinental Ballistic Missiles (ICBMs), enabled states to project their power and deter others from attacking with the threat of retaliation.

During the Cold War era the principle of mutually assured destruction (MAD) deterred the US and USSR from considering a nuclear attack as each had enough weapons that a potential attacker would be punished out of proportion to any advantages from striking first. The Cold War-era nuclear rivalry between the US and USSR led to the notion of strategic stability, an umbrella term that encompasses a range of definitions and ideas. The narrowest understanding of strategic stability refers to the absence of incentives to use nuclear weapons first (crisis stability) and the lack of incentives to build up a nuclear force (arms race stability). It also refers to the existence of a degree of predictability and transparency between two nuclear-armed powers, and the absence of armed conflict between them. The emergence of hypersonics has raised concerns about the continuing efficacy of existing nuclear deterrence and strategic stability, as they could undermine the vital capacity for a credible first and second-strike capability. There is disagreement about their strategic implications however. Some argue that the speed and unpredictability of hypersonic missiles could generate ambiguity around the intended target, potentially triggering strategic miscalculation or unintended escalation. There is also a danger that a lack of clarity over whether a missile is conventionally armed or nuclear could trigger escalation or miscalculation. Cimbala and Lowther (2022, 292–3) suggest that while hypersonics are unlikely in the short term to damage the existing model of deterrence stability based on assured retaliation, if deployed in large number they could alter it by generating uncertainty about the survivability of nuclear retaliatory forces and nuclear command, control and communications systems. Other scholars dispute the contention that hypersonic missiles change either the strategic balance or military capability because states such as China and Russia already possess the ability to launch a massive ballistic missile strike against the US that would overwhelm its missile defence systems (Raitasalo 2019).

Arms Racing? The US, China and Russia

Hypersonic weapons programmes have the potential to fuel an arms racing dynamic. As states gain military advantage, adversaries look for ways to counter it, driving change and innovation, as well as the security dilemma. In order to be able to secure themselves, states must regularly monitor the military capabilities of other states, either allies or adversaries, and respond accordingly, for example, by increasing their own capabilities or engaging in arms control negotiations. The proliferation of weapons can trigger regional arms races, exacerbate existing tensions and increase distrust and hostility. Both China and Russia have made it clear that their hypersonics programmes have been driven in part by the US one (as well as its missile defence system), and vice versa, contributing to a reinforcing effect and arms racing dynamic. The emergence of hypersonic weapons capability challenges the ability of states to manage escalation dynamics, potentially leading to misunderstanding and miscalculation, thereby creating strategic instabilities; the perception of advantage gain is as importance as the reality.

The US is investing in hypersonic capabilities as one of the highest priority critical technology areas to ensure continued battlefield dominance (House Armed Services 2023, 2). It has developed a National Hypersonics Strategy to accelerate the development and delivery of these capabilities, including air, land and sea-launched, intermediate-range, conventionally armed hypersonic weapons. The U.S. Navy's Conventional Prompt Strike (CPS) Program is developing a non-nuclear hypersonic weapon system to provide a prompt, conventional strike capability. There are also several other programs under development, including the Air Force's AGM-183A Air-Launched Rapid Response Weapon (ARRW) and the Army's Long-Range Hypersonic Weapon (Suciu 2021). In addition, the Defense Advanced Research Projects Agency (DARPA) working with the U.S. Air Force has successfully tested its Hypersonic Air-breathing Weapon Concept (HAWC). According to the Department of Defense budget report for FY2025:

> Hypersonic systems will deliver cutting-edge capabilities and strategic options to the Armed Forces to ensure the DoD maintains the ability to deter potential adversaries and defeat aggression whenever necessary. The FY 2025 President's Budget supports developing and demonstrating offensive hypersonic strike weapons, hypersonic defense systems, and critical enablers such as science and technology, workforce development, test and evaluation infrastructure, and industrial base capability and capacity.
>
> (Office of the Under Secretary of Defense 2024, 4–24)

Advocates of the US hypersonics programme argue that, as well as enhancing deterrence, such weapons will ensure that the US retains dominance over its strategic competitors, notably China and Russia. There is also an argument that by investing in hypersonic capabilities, the US is imposing costs on China, which will need to invest time and money in counter-hypersonics. However, as the 2022 National Defense Strategy (US Department of Defense 2022) noted, new technologies such as hypersonics are also complicating escalation dynamics and creating new challenges for strategic stability. The US CPS Program potentially triggered an arms race, as states such as China and Russia sought to counter US advances with their own hypersonics.

China is investing heavily in the development of hypersonic weapons systems, including the Xingkong-2 (Starry-Sky-2), a 'hypersonic vehicle prototype' that was successfully tested in 2018. According to Western analysts, China's military leadership views hypersonic technology as an important element of its regional warfighting strategy, as well as a strategic deterrent (Bernstein & Hancock 2021). Xingkong-2 is believed to have a range of 700–800 km and makes use of an experimental 'waverider' design that uses powered flight after launch, sustaining lift by creating shockwaves. The centrality of hypersonic weapons systems to China's future military capabilities was emphasised by the inclusion of a Dong-Feng-17 (DF-17) HGV at a 2019 military parade led by President Xi Jinping, marking the 70th anniversary of the PRC. The DF-17 is a road-mobile medium range ballistic missile, which carries a HGV with a range of 1,600 km. China has also tested the DF-41 intercontinental ballistic missile, which circumnavigated the globe during a test in July 2021 – prompting a US defence official to compare the event to the start of the space race in the 1950s (Seldin 2021).

Leaked reports from the US Pentagon in 2023 expressed concern that China had tested a longer-range hypersonic missile that was believed to be capable of evading all existing anti-missile defences deployed by the US and its allies (Rogin 2023). The DF-27 is thought to have a range of 5,000–8,000 km, enabling Beijing to strike targets far beyond its traditional area of interest and operation. Chinese military planners have reportedly run war game simulations of a hypersonic strike on a US carrier group in the South China Sea, with 'catastrophic' results for the US CSG (Chen 2023). China's hypersonic capability threatens to shift the strategic balance of power and limit US options for assisting Taiwan in the event of a Chinese invasion.

One of the key concerns over the development of hypersonic weapons systems is their proliferation beyond the US, Russia and China, and fears that smaller states might be able to credibly threaten attacks against the major powers, changing the strategic balance (Speier et al. 2017). Several countries are currently pursuing hypersonic capabilities, including Australia, France, the UK and Japan, which has invested in both an HCM and HGV programme (Yeo 2020). In April 2022, the leaders of the AUKUS defence pact stated a firm commitment to trilateral cooperation on hypersonics and counter-hypersonics (10 Downing Street 2022). India is developing the Brahmos II HCM and North Korea claims to have tested multiple test flights of a hypersonic capability since 2021 (Indian Ministry of Defence 2020).

Russian Hypersonic Weapons Systems and the War Against Ukraine

As discussed above, Russia's war in Ukraine witnessed the first use of hypersonic missiles on the battlefield, when Moscow used its Kh-47M2 Kinzhal missiles to strike Ukrainian targets: on March 19, 2022, the Russian Ministry of Defence claimed to have fired a Kinzhal missile at a munitions store near the town of Deliatyn in the southwest of Ukraine, the first known use of a hypersonic weapon in combat (BBC News 2022).[2] While Russia sought to frame the use of the Kinzhal as being highly significant, the use of hypersonic weapons in Ukraine has not been the game-changer that some thought they might be. The war has seen a return to very low-tech World War Two-era trench warfare and the use of heavy artillery, alongside modern technological developments such as UAVs and hypersonic missiles. Nevertheless, other countries have been paying close attention to these developments: China has reportedly analysed the performance of the Kinzhal on the battlefield, likely drawing lessons for its own hypersonic weapons programme (Goldstein & Waechter 2024).

An article published in a Chinese defence journal sets out a detailed analysis of the Kinzhal's performance in combat and the use of the Su-34 as a launch platform for the missile. It concludes that the use of the hypersonic missile is unlikely to shift the course of the war, noting that after 18 months of war, Russia had few left in its inventory: 'the Kinzhal is expensive to build and is not available in large quantities. It can only be used to attack strategic targets, (Goldstein & Waechter 2024). The author of the Chinese analysis notes a number of defects in both the missile and the Su-34 used to launch it, noting that it was apparently unable to overcome Ukrainian air defences: 'the Ukrainian Ministry of Defence announced that the Ukrainian Air Force shot down a Kinzhal hypersonic missile using the Patriot PAC-3 air defense missile system on May 4, 2023' (Goldstein & Waechter 2024).

There were some critical warnings published in Russian too. In an article published in early February 2022, prior to Russia's invasion of Ukraine, Mikhail Khodarenok, a former colonel who worked within the General Staff's Main Operational Directorate, warned about the dangers of placing too much hope in high-tech weaponry. His analysis provided a clear-eyed assessment of the challenges that might confront an invading Russian force and he warned that stocks of high-precision weapons were not limitless:

> Tsirkon hypersonic missiles are not yet in service. The number of Kalibrs, Kinzhals, Kh-101 (air-launched cruise missiles) and Iskanders is at best measured in hundreds (dozens in the case of the Kinzhal). This arsenal is completely insufficient to destroy a state the size of France and a population of more than 40 million people.
>
> (Khodarenok 2022)

While the use of hypersonic missiles in Ukraine may not have been the silver bullet on the battlefield that Moscow (and others) was hoping for, they have featured heavily in Russian military thinking and strategic planning for a number of years. Russia's development of hypersonic missiles appears to have been driven largely by the US missile defence programme, as well as its development of Prompt Global Strike, which is reported to be capable of delivering precision-guided conventional airstrikes anywhere in the world within an hour, prompting the Russians to develop a similar capability. Russia has had long-running concerns about strategic stability since the end of the Cold War: US missile defence was perceived to have undermined the importance of nuclear weapons, eliminating Russia's strategic parity. However, the debate amongst Russian experts suggests that high precision weapons and hypersonic missiles could go some way to restoring this parity.

A central element of this is the development of HCMs, which are able to circumvent advanced air-defence systems. Equipping the Russian armed forces with high-precision and hypersonic weapons systems was believed to considerably boost its conventional military capabilities and was expected to pose a long-term challenge to potential adversaries. One military theorist has emphasised that victory in future wars will be dependent on an actor achieving superiority over an adversary and gaining (and retaining) the initiative, stressing that in order to achieve this the development of hypersonic missiles will remain a priority direction for Russia.[3]

In his annual address to the Federal Assembly in February 2019, Putin accused the US of pursuing 'absolute military superiority' with their missile defence plans and warned that the Russian response would be 'efficient and effective' (President of Russia 2019). A long-running Russian concern regarding US missile defence centres around the loss of strategic parity with the US; hypersonic missiles are seen as one way to regain parity. Speaking in 2019, Putin set out details of the Avangard hypersonic boost-glide vehicle, which was put into service in late 2019, and the Tsirkon hypersonic missile, saying it could strike targets over 1,000 km away and reach speeds of Mach 9, and warned Russia's enemies to 'calculate the range and speed' of the country's future arms systems (President of Russia 2019). A number of successful test firings of the Tsirkon missile took place from the *Admiral Gorshkov* frigate during 2020 and in October 2021 the Russian Ministry of Defence announced that a Tsirkon missile had been successfully launched from a submarine for the first time (Jonassen 2021).

Putin has described the development of laser, hypersonic and kinetic weapons as a huge breakthrough in Russia's military technology, significantly boosting 'the capacity of the Russian Armed Forces, ensuring a high level of Russian military security for many years, and even decades, to come and it helped strengthen our strategic parity' (President of Russia 2021). This was reflected in the Russian State Armament Programme covering the period 2024 until 2033, which focuses on high-precision weapons, including hypersonic weapons, the introduction of robotic systems, weapons based on new physical principles, electronic warfare equipment and command and control systems based on artificial intelligence (Interfax 2021). These plans are likely to be significantly impeded by the sanctions regime imposed on Russia in the wake of its 2022 invasion of Ukraine, which has drastically limited access to key components such as semi-conductors and microchips.

Conclusions

Do hypersonics represent a revolution in military affairs (RMA) or are they just an evolution? The notion of a fundamental shift in the character and conduct in military operations driven by technological advances was first mooted by Soviet military theorists in the 1980s. Under the leadership of the Chief of the General Staff Marshal Nikolai Ogarkov, they identified a 'military technical revolution' (*voenno-technicheskaya revolyutsiya*), driven by advanced nuclear weapons, the development of long-range precision strike conventional weapons and information technology.[4] The term was a precursor to the idea of a 'revolution in military affairs'

(RMA), which gained prominence in the US in the 1990s, arguing that technology on its own is insufficient to drive major military change; RMAs require operational innovation and changes in doctrine and organisation, alongside new technology (Van Creveld 2010; Metz 1995; Sloan 2002). The information presented above suggests that currently, they are not an RMA, as there has yet to be related change in doctrine and organisation. Furthermore, there are a range of complex technical challenges that need to be resolved in order for hypersonic weapons to become a cost-effective capability. The current significance of hypersonic weapons systems for the character of conflict lies less in their tangible impact on the battlefield and more in terms of what they mean for the global balance of power, their strategic significance, derived from their potential to undermine an adversary's ability to project power. Chinese hypersonics could prevent US aircraft carrier groups operating in the Pacific area, because of their vulnerability to any potential strike; this would undermine the US's ability to project its power on the global stage.

Notes

1 Kahn (1968) wrote: '[O]ne argument against nuclear war is that it may be peculiarly unstable, or volatile, because the tendency for social lethargy to brake violence is reduced nearly to the vanishing point. Preparations for large scale conventional war are painful; for nuclear war, they are not. The restraints on the outbreak of large-scale violence in nuclear war are therefore chiefly, intellectual, ethical, or doctrinal ones'. Herman Kahn, *On Escalation: Metaphors and Scenarios* (New York: Frederick A Praeger Inc., 1968), p. 121.

2 The Kinzhal is a hypersonic aero-ballistic system that is dropped from an aircraft, which accelerates to hypersonic speed using a rocket and then follows a ballistic trajectory.

3 For further details, see Tracey German, *Russia and the Changing Character of Conflict* (Amerhurst: Cambria Press, 2023).

4 For a detailed analysis of the Soviet debates, see Mary C Fitzgerald, 'Marshal Ogarkov and the new Soviet Revolution in Military Affairs', *Research Memorandum*, CRM 87-2, January 1987, Center for Naval Analyses.

References

10 Downing Street. 2022. 'Fact Sheet: Implementation of the Australia-United Kingdom-United States Partnership (AUKUS)' *Policy Paper*, April 5, www.gov.uk/government/publications/implementation-of-the-australia-united-kingdom-united-states-partnership-aukus-fact-sheet/1a1ad299-e2c8-4ef5-82e3-9af0e057fe9a

Acton, James. 2013. 'Silver Bullet? Asking the Right Questions about Conventional Prompt Global Strike', *Carnegie Endowment for International Peace.* https://carnegieendowment.org/research/2014/11/silver-bullet-asking-the-right-questions-about-conventional-prompt-global-strike?lang=en

BBC News. 2022. 'Russia Claims First Use of Hypersonic Kinzhal Missile in Ukraine', March 19, www.bbc.com/news/world-europe-60806151

Bernstein, Paul & Hancock, Dain. 2021. 'China's Hypersonic Weapons', *Georgetown Journal of International Affairs,* January 27. https://gjia.georgetown.edu/2021/01/27/chinas-hypersonic-weapons/

Chen, Stephen. 2023. 'Chinese Scientists War-Game Hypersonic Strike on US Carrier Group in South China Sea', *South China Morning Post*, May 23. www.scmp.com/news/china/science/article/3221495/chinese-scientists-war-game-hypersonic-strike-us-carrier-group-south-china-sea

Cimbala, Stephen J. & Lowther, Adam. 2022. 'Hypersonic Weapons and Nuclear Deterrence', *Comparative Strategy*, 41:3.

Defense Intelligence Agency. 2018. 'Statement for the Record: Worldwide Threat Assessment – 2018', Factsheet, March 6, 2018. www.dia.mil/Articles/Speeches-and-Testimonies/Article/1457815/statement-for-the-record-worldwide-threat-assessment-2018/

Defence Intelligence Ballistic Missile Analysis Committee. 2020. 'Ballistic and Cruise Missile Threat', *National Air and Space Intelligence Centre*.

Goldstein, Lyle & Waechter, Nathan. 2024. 'China Evaluates Russia's Use of Hypersonic 'Daggers' in the Ukraine War' *RAND*, Commentary, January 12. www.rand.org/pubs/commentary/2024/01/china-evaluates-russias-use-of-hypersonic-daggers-in.html#:~:text=Among%20many%20firsts%2C%20Russia's%20invasion,hypersonic%20weapons%20on%20the%20battlefield

House Armed Services Committee. 2023. 'Statement of Vice Admiral Johnny Wolfe, USN Director, Strategic Systems Programs before the Subcommittee on Strategic Forces of the House Armed Services Committee on FY2024 Budget Hearing on US and Adversary Hypersonic Programs', March 10. https://armedservices.house.gov/committee-activity/hearings/strategic-forces-subcommittee-hearing-us-and-adversary-hypersonic-programs

Indian Ministry of Defence. 2020. 'DRDO Successfully Flight Tests Hypersonic Technology Demonstrator Vehicle', Release 1651956, September 7. www.pib.gov.in/PressReleasePage.aspx?PRID=1651956

Interfax. 2021. 'Interview with Yurii Borisov', May 9, www.interfax.ru/interview/764864.

Jonassen, Trine. 2021. 'Russia Test Fires Submarine-Launched Hypersonic Tsirkon Missile for the First Time', *High North News*, October 5. www.highnorthnews.com/en/russia-test-fires-submarine-launched-hypersonic-tsirkon-missile-first-time

Kahn, Herman. 1968. *On Escalation: Metaphors and Scenarios*. New York: Frederick A Praeger Inc.

Khodarenok, Mikhail. 2022. 'Prognozyi krovozhadnyikh politologov', *Nezavisimoe voennoe obozrenie*, February 3. https://nvo.ng.ru/realty/2022-02-03/3_1175_donbass.html

Liebermann, Oren. 2022. 'US Tested Hypersonic Missile in mid-March but Kept It Quiet to Avoid Escalating Tensions with Russia'. *CNN*. April 5. https://edition.cnn.com/2022/04/04/politics/us-hypersonic-missile-test/index.html

Metz, Steven. 1995. *Strategy and the Revolution in Military Affairs: From Theory to Policy*. Collingdale, PA: Diane Publishing.

Oelrich, Ivan. 2020. 'Cool Your Jets: Some Perspective on the Hyping of Hypersonic Weapons'. *Bulletin of Atomic Scientists*. January 1. https://thebulletin.org/premium/2020-01/cool-your-jets-some-perspective-on-the-hyping-of-hypersonic-weapons/

Office of the Under Secretary of Defense (Comptroller)/Chief Financial Officer. 2024. *Defense Budget Overview. United States Department of Defense Fiscal Year 2025*. April 4. https://comptroller.defense.gov/Portals/45/Documents/defbudget/FY2025/FY2025_Budget_Request_Overview_Book.pdf

Office of the Under Secretary of Defense (Comptroller)/Chief Financial Officer. 2023. *Defense Budget Overview. United States Department of Defense Fiscal Year 2024*. March. https://comptroller.defense.gov/Portals/45/Documents/defbudget/FY2024/FY2024_Budget_Request_Overview_Book.pdf

President of Russia. 2019. 'Poslaniye Prezidenta Federal'nomu Sobraniyu', February 20. http://kremlin.ru/events/president/news/59863

President of Russia. 2021. 'Meeting with Defence Ministry leadership and heads of defence industry enterprises', November 3. http://en.kremlin.ru/events/president/transcripts/67061

Raitasalo, Jyri. 2019. 'Hypersonic Weapons Are No Game-Changer', *The National Interest*, January 5, https://nationalinterest.org/blog/buzz/hypersonic-weapons-are-no-game-changer-40632

Reny, Stephen. 2020. 'Nuclear-Armed Hypersonic Weapons and Nuclear Deterrence', *Strategic Studies Quarterly*, 14:4 (Winter): 47–73.

Rogin, Josh. 2023. 'The Most Shocking Intel Leak Reveals New Chinese Military Advances', April 13. www.washingtonpost.com/opinions/2023/04/13/china-hypersonic-missile-intelligence-leak/

Seldin, Jeff. 2021. 'China Hypersonic Test 'Has All of Our Attention', US General Says', VOA News, October 27. www.voanews.com/a/top-us-general-calls-china-s-hypersonic-weapon-test-very-close-to-sputnik-moment/6287945.html

Sloan, E. C. 2002. *Revolution in Military Affairs*. McGill-Queen's Press-MQUP.

Speier, Richard H. et al. 2017. *Hypersonic Missile Proliferation: Hindering the Spread of a New Class of Weapons*, RAND Corporation, www.rand.org/pubs/research_reports/RR2137.html

Stockholm International Peace Research Institute (SIPRI). 2022. 'A Matter of Speed? Understanding Hypersonic Missile Systems'. www.sipri.org/commentary/topical-backgrounder/2022/matter-speed-understanding-hypersonic-missile-systems

Stone, Richard. 2018. 'National Pride Is at Stake', www.science.org/content/article/national-pride-stake-russia-china-united-states-race-build-hypersonic-weapons

Suciu, Peter. 2021. 'US Forging Ahead with Hypersonic Weapons Development Despite Failures', *Clearance Jobs*, August 13, https://news.clearancejobs.com/2021/08/13/u-s-forging-ahead-with-hypersonic-weapons-development-despite-failures/

United Nations Office for Disarmament Affairs. 2019. 'Hypersonic Weapons: A Challenge and Opportunity for Strategic Arms Control', A Study Prepared on the Recommendation of the Secretary-General's Advisory Board on Disarmament Matters. New York.

US Department of Defense. 2022. *2022 National Defense Strategy of the United States of America*, https://media.defense.gov/2022/Oct/27/2003103845/-1/-1/1/2022-NATIONAL-DEFENSE-STRATEGY-NPR-MDR.PDF

van Creveld, Martin. 2010. *Technology and War: From 2000 BC to the Present*. London: Simon and Schuster.

Wilkening, Dean. 2019. 'Hypersonic Weapons and Strategic Stability' *Survival*, 6:5 (October-November): 129–148.

Williams, Ian. 2020. 'Adapting to the Hypersonic Era', CSIS Project on Nuclear Issues, 2 November. https://nuclearnetwork.csis.org/adapting-to-the-hypersonic-era/

Yeo, Mike. 2020. 'Japan Unveils its Hypersonic Weapons Plans', *Defense News,* March 13, www.defensenews.com/industry/techwatch/2020/03/13/japan-unveils-its-hypersonic-weapons-plans/

7 Globalization and Naval Strategy

The Eastward Migration of Sea Power and Its Impact on Maritime Strategy

Sidharth Kaushal

Introduction

Globalization in the 21st century has a character fundamentally different from that of the 19th century. This is not related to the volumes of goods and services traded, which have increased in a relatively linear manner over the last century (with global exports representing 25% of global GDP as opposed to 15% in 1900) (Estaban Ortiz Ospina, 2024). Rather, the two major trends visible over the course of the last decades of the 20th century have been the interaction between different varieties of capitalism with the receding direct economic role of the state in many societies corresponding with the emergence of mixed market models in others and shifts in the locus of industrial production (including in the shipping sector) being driven by the convergence of economies which intentionally run structural surpluses in the pursuit of full employment (such as the PRC) with service oriented economies which run structural deficits (most of the postindustrial world) (Hall, 2001; Prasad, 2016, 50–60). Additionally, it must be noted that in certain ways globalization remains a regionally driven phenomenon. For example, much of the United States trade occurs within the western hemisphere while ASEAN (taken collectively) represents China's largest trading partner (Mission of the PRC to ASEAN, 2024; Office of the United States Trade Representative, 2024).

These trends have had a strategic impact in the maritime domain, where the character of commerce and conflict are intertwined to a degree not always visible in other domains. This impact has been felt in three ways, illustrated by the case of the first island chain and in particular the South China Sea.

First, the ability of the US and PRC to respectively constrain trade have been impacted in different ways. A complex shipping sector in which transactions can occur at sea makes distant blockades increasingly difficult to enforce, eroding the impact of the US' advantage in blue water capabilities. In waters near the littoral, however, when the destination of a vessel is more easily known, this changes. The PRC enjoys considerable advantages here both in the form of its smaller vessels but also its coast guard and paramilitary forces – all of which can be employed to constrain the economies of regional opponents if the PLAN is able to secure sea control.

DOI: 10.4324/9781003520160-7

In the medium term, however, this may change. A major challenge which western states including the US will have to confront in the maritime domain is the bifurcation of maritime power and sea power. States have little control over insurers or private carriers, all of whom may wish to comply with an opposing state's demands. This would appear to be changing with cross-government efforts to control the commercial sector increasingly visible in the US, particularly with respect to critical maritime infrastructure. This in turn is driving a Chinese effort to rely on domestic providers for services such as insurance and the creation of critical infrastructure. A partial deglobalization of the maritime sector can make distant blockades more viable in the medium term along with other forms of economic warfare since each side's trade is forced by state policy onto carriers and infrastructure which are readily identifiable.

Perhaps the most profound shift visible, however, is the change with respect to which states constitute sea powers – a term broader and more encompassing than just naval power (Lambert, 2018 p.5–10). Today, China has a trade to GDP ratio of 37% as opposed to the US' 27% (Macrotrends, 2024). The PRC boasts a larger merchant marine than any other state and has 230 times the US' shipbuilding capacity (Splash 24/7, 2023). Equally, because of the regionalization of trade and the fact that any conflict would be fought in its backyard and not the US' it has far more to lose from commercial disruptions both incidental to a conflict and those which emerge from an opponent's strategy.

For the first time in recent history, then, the US may find itself in the role not of the sea power in a competition but rather that of a relatively insular power with a large navy attempting to offset the advantages of a sea power and disrupt its maritime dependencies. In effect, the US will occupy a role held by competitors such as Germany and the USSR. This will impact every level of force structuring and employment.

The Character of Sea Control

In important respects the character of sea control – the ability to both use the sea for military and commercial purposes and to prevent an opponent from being able to do so – is changing.

For many decades, China's leaders have fretted about the "Malacca dilemma", the notion that the PRC, which is increasingly dependent on external hydrocarbon imports is vulnerable to a distant blockade (Lantegiene, 2008, 143–161). This would seem eminently reasonable and has been the curse of many a nation which had to trade through chokepoints controlled by a rival (for example, Germany during the First World War). Yet what is perhaps most noteworthy about the Malacca dilemma is the degree to which it is a chimera, something which would have not been the case until recently.

The character of international shipping today is sufficiently globalized to make interdiction at reach inherently difficult unless one is also willing to impact every economy within a cordon (including allied ones). Vessels can operate under flags of convenience and their ownership structures often involve multiple stakeholders including registered owners, beneficial owners and ship operators (Shugart, 2024). Vessels can often travel using flags of convenience and can conduct transactions on the high seas, changing their destinations in the process. Unlike years gone by in which the ability to control the activities of a nation's merchant marine went a significant way towards controlling its trade, in a modern context this will not suffice. Even nations which do possess sizeable merchant marines (including China) typically employ state-owned or flagged vessels closer to their home waters, while relying on international shipping further afield. As a consequence, distant blockades will be increasingly difficult to enforce.

It might be argued that the US could enforce something analogous to the navicert system on vessels travelling through the Malacca Strait. Such a system would, however, require the acquiescence of neutral states in a putative conflict at a time when the positions of neutral states will be of particular importance to both belligerents and offending their sensitivities is likely to be avoided. Interdicting vessels which identifiably used Chinese ports might be another option albeit one which would require the cross referencing of imagery and AIS data as well as a legal mechanism to address claims of unlawful interdiction.

An alternative means of achieving sea control might be the denial of insurance to vessels travelling to specific destinations. Reliance on Western insurers has been identified as a strategic risk by Chinese decision-makers in the wake of the oil price cap on Russia (Siqi, 2023). Once more, however, this would be difficult to achieve on a discriminate basis. If insurance was, for example, denied to all vessels operating within a region like the First Island Chainon the basis of war risks, this would impact neutral states. On the other hand if specific legislation was crafted by the US or allied states such as the UK to constrain ship owners voyaging to China, this would require the ability to identify transgressing vessels in large enough numbers to have an economic impact which is challenging for the reasons that interdiction itself is. Moreover, the enabling legislation needed to achieve this would have to be enacted in multiple states in which P&I club insurers operate – something which China could use its considerable economic influence to preclude. Finally, as illustrated by the emergence of Russia's "dark fleet", uninsured vessels could fill the void – although if a nation's trade was forced to rely on such vessels this might simplify the task of interdiction.

The relationship between the state and insurers is also a concern with respect to achieving freedom of action for friendly commerce in the maritime domain. While states themselves can prove resilient to external pressure, the same cannot always be said of actors within the private sector. A recent example of this might be the success of the Houthi movement in driving Western shipping in the Red Sea, a function of the fact that insurers are unwilling to accept the risk of transit (Saul, Cohn, 2024). In the Pacific, China's claims for an extended air defense identification zone in the East China Sea has been repeatedly contested by the air forces of the US and Japan, but private airlines have largely complied with Chinese claims making them a *de facto* reality (CNN, 2013). It remains to be seen how this dynamic will play out with respect to China's coast guard law in the Taiwan Strait or any juridical claims it chooses to enforce in the South China Sea (Heung and Hui, 2024).

The net effect of this is twofold. First, the locus of sea control is moving towards the littoral from the blue water (areas of the open ocean more than several hundred nautical miles from land in which larger navies have historically operated). Part of the reason that China can potentially control commercial activity more easily than the US is by dint of its proximity to key shipping lanes within the First Island Chain and the terminal points for trade (for example, the ports of a competitor state). This persistent presence in the littoral where the destinations of vessels are less ambiguous would appear a precondition for effective sea control over a commercial sector too complex to govern in the blue water or at distant chokepoints.

For the US, in a conflict with China, this would incentivize operating in the littoral particularly since this is where vessels which are unambiguously Chinese are likely to be concentrated (since the vast majority of Chinese flagged and owned shipping operates in the First Island Chain).

However, in order to effectively deny Chinese trade, US vessels would need to be able to operate within the First Island Chain where PRC flagged vessels could be engaged in a more discriminate way. However, this necessarily implies operating in waters contested by a range of anti-access area denial capabilities from longer ranged systems such as the DF-21D anti-ship ballistic missile to anti-ship cruise missiles such as the YJ-18 and diesel electric submarines like the Chinese Type 039 Yuan Class (Kaushal and Markiewicz, 2019 55–60)

This having been said, the ability to constrain shipping closer to an opponent's coastline can be achieved even when a fleet cannot operate freely near it. In the Black Sea, for example, the Russian Navy was able to deny the movement of shipping near Odessa long after the threat posed by Ukrainian Neptune anti-ship cruise missiles and donated Harpoon ASCMs made operating vessels in the northwestern Black Sea unacceptably risky (Kaushal, 2024). This was accomplished primarily by an offensive mining campaign. Mines remain an effective means of enforcing a close blockade near an opponent's coastline and mines which can be remotely activated (for example, the US quickstrike mine) can be laid in peacetime without contravening international law since they pose no risk without having been activated). There are also other means of contesting sea lines of communication in denied waters. For example, the uncrewed surface vessels employed in the Black Sea by Ukraine represent a cheap and, in principle, expendable means of denying access in littoral areas. For many western navies, which have been built around projecting power from the blue water, this will pose a challenge. Many of the capabilities discussed have long been viewed as tools of coastal powers seeking to deny access to a stronger navy rather than as means of achieving sea control per se. This dynamic is certainly visible in the First Island Chain where it is the PLA Navy, which has placed a considerable emphasis on mining. By contrast, despite the US Air Force, Navy and Marine Corps' recent emphasis on stand-in operations and the ability to lay mines such as Quickstrike, this capability has largely languished in the US (and indeed most Western navies). An analogy might be drawn with the way in which navies such as the Royal Navy had to reassess their assumptions about mine warfare during the First World War (Goldrick, 2018).

The second tactical evolution which can be observed in the South China Sea is the importance of proximity and presence and, by extension, numbers of hulls. Tasks such as convoying are inherently resource-intensive. For example, during the Iran-Iraq war the US Navy had to maintain a standing presence of 35 vessels in the Persian Gulf, a commitment which was matched by its European partners (Cordesman, 1990, 70). The challenge of maintaining a standing presence becomes even more difficult if the threats against which convoys are providing protection are both numerous and inexpensive. An example of this has been several standoffs between the PRC and the Philippines near the Scarborough Shoals and the Sierra Madre. The employment of coast guard and maritime militia vessels by the PRC is partially a means of escalation control. However, it also illustrates an inherent challenge in conflict. After the PLA eliminated the threat posed by a smaller state's armed forces in a region like the South China Sea, a blockade could be enforced by coastguard cutters (the largest of which displace 10,000 tons) as well as by militia vessels (Lendon, 2024). Convoys would have to be provided, by contrast, by US Navy vessels which may well be needed elsewhere and which in any case could not easily maintain a tempo of operations needed to achieve such a convoying operation in the way that a large number of paramilitary assets can.

Militarily, then, the impact of globalization on naval combat in the South China Sea has been to shift the locus of sea control operations closer to the littorals because it is only in these areas where an opponent's shipping can be easily identified and controlled. This has in turn created a requirement for mass, which is quite different from the capability by blue water navies built around smaller numbers of high quality vessels and in some cases may well be provided by paramilitary arms of the state.

In the medium term, as will be discussed, a partial deglobalization of the maritime sector may mean that the pendulum may swing back in favor of distant interdiction.

Regionalization, De-Globalization and the (re)Integration of the Government in Sea Control

In principle, sea control has always been a "whole of government" task. For much of its history, the British Admiralty employed more civil servants than military officials, a reflection of the fact that the maintenance of sea power involved more than the navy itself (Lambert, 2012). For example, Britain's wartime blockade of Germany was made possible in part by civilian inspectors (some posted in neutral states' harbors) as well as courts of law to adjudicate contesting claims about vessels' destinations.

In recent decades there has been something of a bifurcation between maritime activity and naval activity in many Western states. This is, however, arguably untenable. In the Red Sea the consequences of this bifurcation have been visible in the fact that naval vessels have run maritime protection operations which have, despite tactical effectiveness, not restored shipping to anything like pre-conflict levels since insurance costs have continued to rise. In the East China Sea a Chinese Air Defence Identification Zone, which has been repeatedly contested by US and Japanese aircraft nonetheless sees compliance (albeit with the blessing of the US government) from civilian airliners (CNN, 2013).

The mismatch between naval and government activity is noted by the PLA and would likely form a major part of its approach to competition within the First Island Chain. For example, Chinese doctrinal literature regarding blockades places a particular emphasis on the concept of a "comprehensive blockade" which would see Chinese diplomats engage in outreach to both states and commercial actors in the states to which they are posted to secure compliance with blockades (Kaushal, 2022). For nations with a more bifurcated understanding of naval power and maritime power this could pose considerable challenges – convoying, after all, makes little sense when there is nothing to convoy.

However, the dynamics of Sino-American competition in the South China Sea have seen a growing emphasis on cross-government coordination within the US in recent years. The most obvious instance of this was the US effort to exclude Chinese companies from the consortium intended to build the SEA-ME-WE-6 cable linking Europe and Asia. The effort involved a combination of diplomatic pressure (led by the State Department) the threat of sanctions (in which the Treasury would have played a role) and preferential loans for competitors organized by the US Export Import Bank. This coordination is however *ad hoc* and issue-specific at the moment. It would appear more likely than not that the dynamics of strategic competition will make it more routine. One consequence of this will be a partial deglobalization of the economy and by extension the maritime system. This is visible in the arena of undersea infrastructure where Chinese companies are, backed by the state, increasingly setting up parallel cables to offset the effect of their exclusion from many international competitions (Brock, 2023). It is increasingly visible in the world of maritime insurance where the already rapid growth of the Chinese maritime insurance market will likely be further catalyzed by lessons learned from the Western oil cap on Russia (South China Morning Post, 2023; Ostler, 2021).

The dynamics of a partial deglobalization will be reinforced by one other – regionalization. As a standard gravity model would predict, most exchanges in goods and services remain regionally oriented. Southeast Asia, for example, does far more trade with China than any other state and trade with ASEAN as a bloc is China's largest export relationship (Perez Gill, 2023). One consequence of this is that disruptions to regional trade within the South China Sea would likely, over time, have a far greater impact on the PRC than on the United States. Indeed by some estimates wartime disruptions to trade within East Asia could cost the PRC 35% of its GDP in a year of conflict (as opposed to 17% for the United States) (Gompert, 2016, 48).

While the previous section of this chapter suggested that globalization would necessarily force navies into the littorals of the South China Sea (among other seas) a long term trend, which sees the role of the state reasserted and in which regional trade comes to matter more than global trade, could have the opposite effect. If, for example, maritime infrastructure becomes neatly bifurcated along state lines, the incentives for sabotage both for wartime impact and to shape the post war commercial order become greater for both competitors, particularly as targeting infrastructure involves fewer problems of entanglement (as is the case when both parties use the same infrastructure). Similarly, long term shifts in the global insurance industry would make it easier to seize or redirect vessels via a distant blockade based on the insurance they are employing. Finally, if the US deems costs imposed on friendly or neutral states within the cordon acceptable (as Britain did during the First World War) a distant blockade may have a far greater impact on the PRC than the US.

If the present era, perhaps a high point of globalization, is one in which the sheer complexity of the maritime economy and the removal of Western states from many aspects of it make effective sea control difficult to achieve except in littoral areas, the steps which states take to achieve greater agency and the partial economic bifurcation this drives could, along with existing trends towards regionalization, make the restoration of effective sea control by blue water navies more viable in the medium term.

Industrial Capacity – From Using Sea Power to Overcoming It

A final trend worthy of note in the maritime domain is the shift in the structure of global shipping. As of this writing, East Asia accounts for roughly 85% of global ship construction with China, Japan and South Korea making up the lions share of this figure (Looman and Kang, 2024). China in particular has seen its role in this sector propelled by a combination of lower input costs, state support for shipbuilding and the requirement to find an outlet for excess capacity in areas such as the steel sector account for this. China's excess steel capacity in 2016 was equivalent to the steel production capacity of the UK, France and Germany (Beckley, 2016, 150). While this excess capacity might be commercially inefficient, the need to channel it towards productive functions has conferred on the PRC considerable strategic advantages since strategic success is often less a matter of commercial efficiency than the ability to maintain capacity and redundancy.

China's major shipyards benefit from large numbers of both civilian and military orders, with the absence of a clear bifurcation between civilian and military shipbuilding (as is often the case in Western nations) meaning that skills within the workforce can be maintained and employed without military shipbuilding having to account for all of a given shipyard's demand. Instead, military shipbuilding has often played a counter-cyclical role – compensating for shortfalls in civilian demand in lean times and benefiting from considerable civilian shipbuilding demand (which keeps shipyards active without drawing on defense budgets) in others (Predd et al., 2024, 5).

This has led to the emergence of massive facilities such as Jiangnan shipyard which produce both civilian and military vessels at a scale which the US would struggle to match (CSIS, 2018). For example, in 2023 the PLAN added 15 new surface combatants to the US Navy's 2 and the aggregate shipbuilding capacity of the PRC is 230 times that of the US (Williams, 2024).

For the US Navy, despite its qualitative advantages, this would pose significant challenges in a theatre such as the South China Sea. First, attrition at sea is by definition more consequential for a more distant state that cannot rely on ground based airbases and missiles for a portion of its firepower at sea. In effect, the US Navy would begin any fight with less slack capacity than the PLAN even on even terms. The differential in shipbuilding capacity would only compound this challenge. There is some evidence to suggest that in conflicts between technologically comparable fleets, larger fleets have consistently prevailed even when a smaller fleet enjoyed early technological advantages (Tangredi, 2023).

There are, however, cases where this advantage on the ocean surface has not mattered. For example, early in the war in the Pacific the Imperial Japanese Navy enjoyed considerable advantages in maritime East Asia but struggled to contend with the US submarine threat, a challenge also faced by the Royal Navy. The US' enduring advantages in the quietness of its nuclear attack submarines and the maturity of its anti-submarine warfare capabilities vis the PLAN might prove an asymmetrical offset. Additionally, for many of the aims it might seek to achieve in East Asia the PLAN requires sea control whereas for the US, sea denial may well suffice (Colby, 2022). This aim can be achieved with submarines but also with other standoff capabilities such as missiles like the long range anti-ship missile launched from bombers like the B1B (NAVAIR, 2017). That China's surface fleet can control an area, then, is not tantamount to victory.

Finally, with its greater dependence on both world trade and critically regional trade, China stands to lose far more from maritime disruptions within the First Island Chain than the US, which can maintain access to many of its core markets even if a conflict breaks out in East Asia. While mutually damaging, the conflict will be more damaging to a Chinese economy which is, paradoxically, more dependent on trade than the US despite the PRCs quasi-statist model.

This does place the US in a somewhat unfamiliar position, however, of playing a role analogous to the one played by states such as Germany and the USSR during the World Wars and the Cold War. Historically the dominant maritime player with the imperative to control the surface of the ocean, the US Navy had to consider how it would secure and then employ sea control. As a nation that is now, in some regards, a post-sea power state, the US may find itself employing tools such as bomber launched standoff, submarines and commerce disruption which would be familiar to any student of the USSR's naval planning during the Cold War. A post-sea power state confronting a new sea power in the South China Sea may, then, have to rely on an unfamiliar toolkit.

Alternatively, the importance of regional naval powers may grow as they come to represent a means of closing the gap between the number of hulls available to the US and the naval capacity at its disposal. This could occur in several ways. First, the US may find itself reliant on shipyards in Allied nations such as South Korea in order to construct vessels (Nemeth). Equally, the growing navies of regional powers such as India will likely play an ever greater role in enabling burden-shifting in regions such as the Indian Ocean in order to concentrate US capabilities near the First and Second Island Chains (Pattatathunaduvil, 2024).

Conclusion

The impact of globalization on the character of naval warfare in regions such as the First Island Chain will be profound. At the heart of this is the fact that globalization has changed both the character of sea control and has brought new sea powers into being.

Sea control will in the short term increasingly depend on the ability to disrupt an opponent's trade close to his shores and to coordinate the actions of players beyond the navy (and indeed government), including insurers, in order to ensure the movement of goods to friendly states. In certain ways, this may drive a renaissance of the classical approach to marshalling a state's sea power. In the medium term, however, a different trend might be observed. The bifurcation of the global maritime sector in areas such as insurance, shipbuilding and the construction of critical infrastructure may allow more discriminate targeting of an opponent's commerce at reach, incentivizing modes of warfare such as commerce raiding and distant blockades. For China, this will have considerable negative ramifications.

Perhaps most profound, however, is the shift in the identities of global sea powers. China, historically a continental power, meets any reasonable definition of a sea power to a greater degree than the United States. Its trade to GDP ratio is considerably greater than the US as is its merchant marine and its shipbuilding capacity. However, the US enjoys the advantage of geography, along with enduring advantages in areas such as submarine quietness. Moreover, the very fact that China's trade has to pass through chokepoints and that the PRC is more vulnerable to global economic disruptions than the US is an asymmetrical advantage. Ironically, then, the US finds itself in the position typically inhabited by its competitors as it will have to contemplate not the question of how to use sea power against an insular continental state but how to disrupt and deter a sea power as a (comparatively) insular state. As Western admirals and planners look for answers, then, examining the successes and failures of "the other side" of major 20th-century conflicts may prove more instructive than the history of western navies in this era.

References

Beckley, Michal. *Unrivalled: Why the US Will Remain the World's Sole Superpower*. Ithaca, NY: Cornell University Press, 2016.

Brock, Joe. 2023. "Exclusive: China Plans $500 Million Subsea Internet Cable to Rival US Backed Project." *Reuters*, April 6, 2023. https://www.reuters.com/world/china/china-plans-500-mln-subsea-internet-cable-rival-us-backed-project-2023-04-06/.

Colby, Elbridge. 2022. *Strategy of Denial: American Defense in an Age of Great Power Conflict*. New Haven, CT: Yale University Press.

Cordesman, Anthony. 1990. *The Lessons of Modern War – Volume II – The Iran-Iraq War – Chapter 14: The Tanker War And The Lessons Of Naval Conflict*. Washington, DC: Centre for Strategic and International Studies.

CNN. 2013. "US Airlines Comply With China's Demand For Notice of Flights Through Zone." *CNN*, November 30, 2013. https://edition.cnn.com/2013/11/30/world/asia/china-japan-us-tensions/index.html.

CSIS (Center for Strategic and International Studies). 2018. "Analysis of Jiangnan Shipyard." https://chinapower.csis.org/analysis-jiangnan-shipyard/.

Gil Perez, Javier, et al. 2023. "China's Bid for Leadership in Southeast Asia." *Barçelona Centre for International Affairs*, November 2023. https://www.cidob.org/en/publications/chinas-bid-leadership-southeast-asia.

Goldrick, James. 2018. "Antiaccess for Sea Control: The British Mining Campaign in World War I." *US Naval Institute Proceedings*, October 2018. https://www.usni.org/magazines/naval-history-magazine/2018/october/antiaccess-sea-control-british-mining-campaign-world.

Gompert, David C., Astrid Stuth Cevallos, and Cristina L. Garafola. 2016. *War with China: Thinking Through the Unthinkable*. Santa Monica, CA: RAND Corporation.

Hall, Peter A., and David Soskice, eds. 2001. *Varieties of Capitalism: The Institutional Foundations of Comparative Advantage*. Oxford: Oxford University Press.

Heung, Chloe, and Karen Hui. 2024. "China's New Coast Guard Regulations Up the Ante in Tense South China Sea." *Asia Pacific Foundation of Canada*, July 4, 2024. https://www.asiapacific.ca/publication/chinas-new-coast-guard-regulations-in-south-china-seas.

Kaushal, Sidharth. 2024. "Lessons for the Royal Navy's Future Operations from the Black and Red Sea." *RUSI*, July 6, 2024. https://www.rusi.org/explore-our-research/publications/commentary/lessons-royal-navys-future-operations-black-and-red-sea.

———. 2022. "The PLA Approach to Blockade Operations." *RUSI*, November 3, 2022. https://www.rusi.org/explore-our-research/publications/rusi-defence-systems/pla-approach-blockade-operations.

Kaushal, Sidharth, and Magdalena Markiewicz. 2019. "Crossing the River by Feeling the Stones: The Trajectory of China's Maritime Transformation." *Royal United Services Institute*.

Kang, Min Joo, and Richard Looman. 2024. "Asia's Shipbuilding Renaissance: Record Orders and Rising Prices." *ING*, December 16, 2024. https://think.ing.com/articles/asia-shipbuilding-renaissance/.

Lambert, Andrew. 2018. *Seapower States: Continental Empires, and the Conflict That Made the Modern World*. New Haven, CT: Yale University Press.

Lambert, Nicholas. 2012. *Planning Armageddon: British Economic Warfare and the First World War*. Cambridge, MA: Harvard University Press.

Lanteigne, Marc. 2008. "China's Maritime Security and the 'Malacca Dilemma.'" *Asian Security* 4(2): 143–161. https://doi.org/10.1080/14799850802006555.

Lendon, Brad. 2024. "What Is China's 'Monster' Coast Guard Ship and Why Is the Philippines Spooked by It?" *CNN*, July 8, 2024. https://edition.cnn.com/2024/07/08/asia/china-monster-coast-guard-ship-philippines-intl-hnk-ml/index.html.

Macrotrends. 2024. "China Trade to GDP Ratio 1960–2024." https://www.macrotrends.net/global-metrics/countries/chn/china/trade-gdp-ratio.

———. 2024. "US Trade to GDP Ratio 1960–2024." https://www.macrotrends.net/global-metrics/countries/usa/united-states/trade-gdp-ratio.

Mission of the PRC to ASEAN. 2024. "Looking into the Future Seeking Common Development." http://asean.china-mission.gov.cn.

NAVAIR. 2017. "Navy Completes First LRASM Free Flight from B-1B Lancer." https://www.navair.navy.mil/node/25561.

Nemeth, Bence. n.d. "South Korea's Role in Reviving the U.S. Navy." *Korea Institute For Maritime Strategy*. https://en.kims.or.kr/issubrief/kims-periscope/peri365/.

Office of the United States Trade Representative. 2024. "Western Hemisphere." https://ustr.gov/countries-regions/americas.

Ortiz-Ospina, Esteban. n.d. "Trade and Globalization." *Our World in Data*. https://ourworldindata.org/trade-and-globalization.

Osler, David. 2021. "Chinese Hull Market Overtakes London's." *Lloyd's List*, September 6, 2021. https://www.lloydslist.com/LL1138098.

Pattatathunaduvil, Pranav. 2024. "Ways to Strengthen the U.S.-Indian Naval Partnership." *US Naval Institute Proceedings*, May 2024. https://www.usni.org/magazines/proceedings/2024/may/ways-strengthen-us-indian-naval-partnership.

Predd, Joel, et al. 2024. *PRC Shipbuilding: Naval and Commercial*. Santa Monica, CA: RAND Corporation.

Prasad, Eswar. 2016. *The Dollar Trap: How the US Dollar Tightened Its Grip on Global Finance*. Ithaca, NY: Cornell University Press.

Saul, Jonathan, and Carolyn Cohn. 2024. "Red Sea Insurance Costs Soar as Houthi Shipping Threats Loom, Sources Say." *Reuters*, September 19, 2024. https://www.reuters.com/world/middle-east/red-sea-insurance-costs-soar-houthi-shipping-threats-loom-sources-say-2024-09-19/.

Shugart, Thomas. 2024. "There Are No Magic Beans: Easy Options to Deter China Militarily Do Not Exist." *War on the Rocks*, August 23, 2024. https://warontherocks.com/2024/08/there-are-no-magic-beans-easy-options-to-deter-china-militarily-do-not-exist/.

Siqi, Ji. 2023. "China Takes Lessons from Russia, Out to Fix Maritime Insurance 'Weakness' after Ukraine War." *South China Morning Post,* April 17, 2023. https://www.scmp.com/economy/china-economy/article/3217243/china-take-lessons-russia-out-fix-maritime-insurance-weakness-after-ukraine-war.

Splash 24/7. 2023. "China Beats Greece to Become the World's Largest Shipowning Nation by Gross Tonnage." https://splash247.com/china-beats-greece-to-become-the-worlds-largest-shipowning-nation-by-gross-tonnage/.

Tangredi, Sam. 2023. "Bigger Fleets Win." *USNI Proceedings*, January 2023. https://www.usni.org/magazines/proceedings/2023/january/bigger-fleets-win.

Williams, Lauren C. 2024. "China Is Winning the Shipbuilding Numbers Game—and That's a Problem, INDOPACOM Nom Says." *Defense One*, February 4, 2024. https://www.defenseone.com/threats/2024/02/china-winning-shipbuilding-numbers-gameand-s-problem-indopacom-nom-says/393904/.

8 Towards a Maritime Strategy for the Second Revolution in Military Affairs

James Henry Bergeron

Introduction

Maritime strategy is back in vogue and it is not difficult to see why. In the decades after the fall of the Berlin Wall, the seas were viewed by navalists, and indeed were, little more than a highway to move warships for power projection ashore in crisis response, or to the littorals for maritime security or counter-terrorist operations. That strategic re-envisioning from collective defence and the protection of sea lines of communication (SLOCs) to crisis intervention was first captured in the US Navy's " ... From the Sea" (Department of the Navy, 1992) and "Forward ... From the Sea" (Department of the Navy, 1994), followed by "A Cooperative Strategy for 21st Century Sea Power" (Department of the Navy, 2007). The US strategies were paralleled by the United Kingdom's "Options for Change" (House of Commons, 1990: 468–86), the 1998 Strategic Defence Review (UK Ministry of Defence, 1998), and 2010 Strategic Defence and Security Review (HM Government, 2010), as well as by NATO's 2011 Alliance Maritime Strategy (NATO, 2011). Together, these statements set the tone for maritime strategy in the post-Cold War era.

Although all of these works emphasized the preservation of potent warfighting capabilities, albeit at smaller scale and shorter endurance, a core assumption of those years was the likelihood of maritime and air supremacy over potential adversaries. This was largely assured, thanks to the precision weapons systems developed in the 1980s and 1990s, a technological leap forward twinned with a new way of war, and labelled – not without controversy (Biddle, 1996) – as the beginning of a revolution in military affairs (RMA) that had its first spectacular success in the Gulf War of 1991 (Dombrowski and Ross 2008, 13). There was also the lack of a conventional peer competitor. The Russian Federation Navy (RFN) was convalescing from its collapse in the early 1990s, while China's People's Liberation Army (Navy) was only taking its first steps as a blue-water fleet in counter-piracy operations off the Horn of Africa. Although navies, come budget time, routinely stressed the overwhelming importance of the seas for global trade, mobility, and communications, there was no serious threat, with the partial exception of piracy, to that public good.

DOI: 10.4324/9781003520160-8

On land, the story was different. The superiority of US and Allied strike capabilities based on network-centric warfare led to asymmetric responses from terrorist and insurgent groups from Al Qaeda's global terror campaign to the militant insurgencies in Iraq and Afghanistan. These underscored a key element of any such revolution: that it changes the options and strategies of *all* sides, in an effort to maximize leverage while minimizing vulnerability.

Further, beginning in the mid-2010s, the overwhelming superiority of allied naval forces began to change. Russia created a new, lighter fleet of destroyers, frigates, and diesel submarines armed with Kalibr land attack cruise missiles, and began using those forces in Syria (Bogdavov and Kramnik, 2018). The long-delayed Yasen-M Class guided missile nuclear attack submarine (SSGN), advanced and quiet, began to be commissioned and deployed to sea (Kaushal et al., 2021). The PLA(N) became the world's largest navy in terms of platforms, albeit essentially untried in combat. Russian research ships began an era of deep-sea surveillance and potential disruption of undersea infrastructure (Kaushal, 2023; Sutton, 2021).

Then came the full-scale war on Ukraine in February 2022. Commercial shipping and ports were attacked for the first time in Europe since World War Two, and a blockade on the export of grain was imposed. Ukraine responded with its brilliant use of land-based anti-ship missiles and autonomous undersea vehicles (USVs) against a RFN that was unprepared to challenge them (Sutton, 2024; Shuster, 2024). The result was to sweep the Russian Black Sea Fleet from the western Black Sea in the first half of 2024 and attack them in their own naval base at Sevastopol. It was a devastating demonstration of new technology, but even more of human ingenuity in the rapid resourcing, deployment, and tactical development of weapons systems made largely from home grown, off-the-shelf commercial components (Shuster, 2024).

The drone transformation of the land war was even more remarkable, with more than one million drones, often hand-held attack vehicles, fielded to date by each side. In October 2023, President Zelensky announced that Ukraine has the capacity to produce four million drones of more than 100 types annually, a target Russia is expected to match (Franke, 2025). Infantry battles are now largely drone-based, as are anti-armour attacks and, to a degree, air defence (Sommerville, 2024). The conduct of land warfare was rapidly transformed, far faster than Western militaries expected or were prepared for. This has implications not only for conventional army forces and the tactics of major Allied land powers, but for the role of sea power as domain advantages shift and churn.

In parallel, and with Russian and Iranian help, from October 2023, the Yemeni Houthis shifted their land-based drone campaigns against the recognized Yemen government and Saudi Arabia to begin a missile and drone campaign against merchant shipping in the Red Sea and Gulf of Aden. This campaign diverted some sixty percent of global trade through the Suez Canal to the long southern route around Africa, inducing supply chain shocks, a rising of consumer prices, and inflation (Berman, 2024; Meade, 2024; Lawford, 2024). In July 2024, a Houthi missile travelled 2600 miles to strike a target in Tel Aviv, demonstrating their increasing range and possibility of penetrating defences (Nevola and d'Hauthuille,

2024). At the time of writing, despite a Houthi declaration that it will cease most attacks following the Gaza ceasefire agreement of January 2025, the new pattern of trade appears long-term. But even if the Gaza ceasefire holds, the Houthis have demonstrated the ability of a well-trained and supplied militia to disrupt global shipping, with the circumnavigation of Africa remaining the one safe route (at present) for more than half of East–West global trade.

Although Allied navies have been successful in countering attacks on commercial shipping and warships in the Red Sea and Gulf of Aden, the Houthi offensive highlighted the deep asymmetry of using exquisitely advanced, extremely expensive air defence missiles against cheaper missiles and drones (Slayton, 2024). In late 2024, increasing use of non-kinetic responses by US, UK, and European warships were reported (discussed below), indicating a learning curve and countermove with new technology of their own to the new challenge beyond conventional tactics.

The Ukraine War and Houthi strikes do not stand in isolation. In 2010, three states fielded drones for military purposes. By 2023, that number rose to 40, with autonomous systems employed in 34 conflicts worldwide (Nevola and d'Hauthuille, 2024). However, the Black Sea and Red Sea conflicts shifted a paradigm of conventional conflict that was already under heavy strain. This heralded the dawn of a second RMA (to be referred to henceforth as 2RMA), as the Gulf War sparked the notion of a RMA for the 1990s. Cheap and plentiful autonomous systems are not the only elements in a wave of emerging technologies. Advances in cyber warfare, space technology, quantum engineering, artificial intelligence (AI), and hypersonics will also shape the military instrument of power for great powers with large defence budgets and industries. But the autonomous successes in the field in 2023–24, the shift in availability and affordability to mid-sized powers and non-state actors, and the following rapid innovation in counter-drone and counter-counter drone systems and tactics by all sides to conflict arguably define a new era.

New technology and innovative employment does not count for everything, of course, as the scale and readiness of conventional naval forces are equally important. In the Pacific, the PLA(N) routinely conducts massive naval encirclement exercises around Taiwan and continually threatens to incorporate it into the People's Republic by force if necessary. Lloyds has estimated that a major global conflict that disrupted the key trade routes or isolated Taiwan could cost global GDP between 7.4 and 55 trillion US dollars (Cohn, 2024). The year 2024 saw unambiguous concertation between the "Gang of Four" authoritarian regimes of Russia, China, Iran, and North Korea (Chivvis and Keating, 2024). Russia and Iran, at least, seem to have the harassment or disruption of global trade and communications within their sights, while China, more dependent on international commerce, seems to be willing to risk the threat of regional disruption to achieve its strategic ambitions in the Pacific.

The NATO Allies continue to possess a naval force substantially superior to the RFN, and the battle-experienced USN is more than comparable to the PLA(N). But naval power is not only, or even mostly, about defeating an enemy fleet, ship for ship. It is about contributing to, or *threatening*, the defeat of the adversary itself, upending their theory of victory, their *strategy*. That is the essence of a strategy of deterrence and the maritime contribution to it.

Further, Allied navies are challenged in places by aging fleets, maintenance delays and cost overruns, pressure on defence budgets, and recruitment and retention. There is public debate, in the United States but also elsewhere, on the balance to be struck between restoring a somewhat old-school fleet to readiness, or to shift investment and risk towards radical new 2RMA technologies and systems, a tension captured well in the US Chief of Naval Operations *Navigation Plan for America's Warfighting Navy 2024* (United States Navy, 2024) – arguably the first major maritime strategy (*sans* the name) of the post-Cold War era.

The explosion of hybrid and grey-zone warfare in the 2020s also speaks to the balance to be struck between technological innovation and the need for platforms. The year 2024 witnessed a string of incidents damaging critical undersea infrastructure (CUI) in the Baltic, caused by repeated incidences of dragging anchors over cables or pipelines (Grady, 2024; Kauranen et al., 2024). Whether ordered by the Kremlin or cases of criminal negligence remains undecided at the time of writing. Regardless of intent, navies will need ships, patrol aircraft, and drones, in large numbers, to monitor and respond to such very low-technology threats to CUI. That puts further pressure on fielding a ready fleet now.

Strategy's Clausewitz moment comes where, in a resource-constrained environment (and it is always constrained), hard choices must be made to create the means that can credibly deliver the ends of policy. That conversation is ongoing, particularly in the United States, but also in several Allied states, and in NATO. The analysis required to inform the specifics of such a strategy is a major undertaking, beyond the scope of this contribution. Rather, this chapter hopes to provide grounding and context to that conversation by going back to basics on what a maritime strategy is, what it requires, and in particular how the geo-political ends of national and Alliance strategy relate to the debate over ways and means of sea power triggered by the 2RMA phenomenon.

What Is a Maritime Strategy?

What does it mean to have a "maritime strategy"? The term has been used in very different ways. The most typical contemporary meaning is of a formal document, usually unclassified, designed to help deter adversaries, assure allies, and convince publics and parliaments of the wisdom of investing in their navy. Alternatively, a maritime strategy can be considered as just one of many component strategies in a comprehensive joint warfighting plan. At the level of grand strategy, states that have maritime concerns or advantages at the core of their existence can develop national strategies that are essentially maritime, as Lambert noted in the case of Seapower States (Lambert, 2018).

Maritime strategy also has another aspect, at least in the community of nations that arose as maritime trading states and relatively open societies (Padfield, 1999). A maritime strategy designs a fleet, but it also defines a nation, or an alliance. More than almost any other warfare domain, a maritime strategy, if serious, is also about designing the kind of country, even the kind of polity, the authors want to mirror or, as with Mahan, transform itself into (Mahan, 1890; O'Connell, 1993). Ethics, political order, political economy, the openness of a society, and a people's view of themselves in the cosmos, are all implicated by the kind of maritime strategy maritime trading states embrace.

At the most basic level, an effective maritime strategy must engage the critical geopolitical and strategic issues of the era as its authors understand them. In times of change, it needs to be controversial, relevant, and creative. Its fortunes will be tied – and should be tied – to the outcome of political debates over those primary issues in capitals and alliances. The alternative is to draft a "safe maritime strategy" in more service-oriented, general, and abstract terms that could arguably remain applicable regardless of major policy changes or shifts in the geopolitical global order or domestic politics at home. Although it might "make the maritime case" in general terms, there is a danger that it will be seen as "nice to have" but ultimately irrelevant to the issues that matter. It would also likely prevaricate on precisely those hard choices that must be made. Such a strategy could quickly gather dust, lacking the political and economic momentum to support it.

A successful strategy therefore needs to connect ends, ways, and means that speak to the concerns of vital stakeholders, while retaining a true and effective strategic vision grounded in naval realities. These include political and military leaders, as well as parliaments and financial markets, which will be expected to foot the bill. This puts a premium on the creative application of means to ends, as well as on the potential redefinition of policy goals due to constrained capabilities or unaffordable costs (e.g. the British decision Pre-World War I to cede maritime dominance in the North Atlantic and Pacific to others in order to concentrate forces in the European theatre). Finally, the public: a viable maritime strategy needs to "ring true" to the citizen; it must be capable of justification and support by national governments before their own publics, in the glare of the media.

A major consideration in strategic planning is time scale, and this is especially important during periods of rapid technological change. A maritime strategy is crafted in the context of near-immediate operational needs, the assumed intent of adversaries in the medium term, the R&D, and development cycles of new technology which is now measured in months or a few years, and the longer budgeting and shipbuilding processes of Western states which can span from several years to decades. It needs to steer a middle course between security "flavours of the month," as well as strategic horizons that are so broad or distant as to lose practical meaning. A stand needs to be taken on which problems are expected to endure, which are expected to grow, and which are local or ephemeral over a five-to-thirty-year time span.

Economics teaches us that, in the short term, almost all costs are fixed, i.e. the decision-maker is dealing with sunk costs, fixed investments, and limited flexibility in the creation of new capabilities. In the short term, the primary "strategic" focus is *a fortiori* tactical, how to best employ resources at hand, exemplified in the NAVPLAN 2024 efforts to ready the US Navy for potential conflict by 2027, a mere two years away (United States Navy, 2024, 6). In the longer term, however, most costs become variable, capabilities open to investment and redefinition. The hard reality is that action must be taken for both the short term and the long term now, not later, although there is time to adjust the longer-term strategy. This presents a dilemma when a technological generation is so short. A viable maritime strategy for the 2RMA era needs to address the current technological revolution, which is a debatable known, as well as to make an educated guess on the one likely to follow in five to ten years' time, a Rumsfeld "known-unknown" (US Department of Defence, 2002). Over the typical life of a warship, over half may be spent in the land of R&D "unknown-unknowns" as seen at the time of its design.

Ultimately, strategy requires choices, it does not provide them. As noted, the essential assumption of all strategic thought is limited means, requiring a political choice between alternative geopolitical goals, and a creative military and whole of government choice between alternative ways of applying constrained means towards chosen ends. Assertions that the navy or the Alliance "must" be able to undertake all missions in all places concurrently, to operate across "the full spectrum of maritime tasks," etc., tends to empty any maritime strategy of its intellectual bite since there are no hard choices to be made.

An effective strategy, therefore, should be based on primary political objectives and values. It should be associated with a specific, identified problem and defined set of actors that the strategy is intended to influence. It should contain a course of action that chooses between rival, not unreasonable solutions; a hypothesis of adversarial reactions to that choice; an argument of comparative strategic advantage and hypothesis that adversaries will respond to that advantage in desired ways. In short, a theory of victory and an articulation of gains.

Framing a New Maritime Strategy: The Ends and Ways of Policy

The Western maritime strategies of 1989–2016 were based at the high end on maintenance of nuclear deterrence (to maintain the mutual deterrent equilibrium), conventional superiority, and the pursuit of strategic stability. The latter was primarily pursued via global maritime power projection by the United States and to a degree by the United Kingdom, France, and a few other European naval powers. This power projection strategy was focused on crisis management operations against a set of conventionally inferior, non-nuclear adversaries, based mostly on the RMA recipe of precision strike, land attack cruise missiles, data links, and instantaneous communications. This delivered air and maritime supremacy, although operational success still remained challenging in places such as Afghanistan due to challenging geography and the asymmetric, insurgent nature of the conflict.

The second concern of that era (and today) was maritime security, whether as part of the global war on terror, counter-piracy operations, counter-proliferation, or attempts to stop illegal migrant trafficking. In the post-Cold War era, maritime security became the major naval preoccupation for NATO, from Operation Active Endeavour (OAE) to Operation Ocean Shield in the Horn of Africa and the present successor of OAE in the Mediterranean, Operation Sea Guardian (Bergeron, 2024b). It was explicitly incorporated as a major role for the Alliance in the Alliance Maritime Strategy of 2011 (Bergeron, 2024b, 176–77). This form of maritime security was built around frigates, corvettes, and maritime patrol aircraft. Its focus was maritime situational awareness, literally human eyes-on and within radar range, while the main means of response, at least in theory, was maritime interdiction or boarding operations. It was not a strategy of sea control *per se*, or the defence of SLOC against a comprehensive threat, instead being closer to constabulary operations.

The changes, in politics, geostrategic competition, and technology, of the post-Cold War era were summarized in the first section of this chapter. This begs the question: What a maritime strategy for the 2RMA should be, in this new era and envisaging the one to follow?

The first question is: to what extent 2RMA changes the essentials of strategy? I suggest that the changes driven by 2RMA exist primarily in the realm, not of ends or even ways, but of means. However, it was always misleading to assume that the three aspects of strategy do not relate to each other dynamically. Changes of means due to economy and technology, and the relative advantage or vulnerability resulting, impact all facets of strategy through the opportunities and costs they impose on choosing geopolitical objectives.

To begin with those objectives: at the most general level, the ends of Western national and Alliance policy remain relatively constant: to defend national independence, preserve democratic governance, support economic prosperity, and maintain global strategic stability. At the time of writing, articulating the near-term geopolitical threat environment is difficult, but more certain, for long-term trends. Over the next year, the Russia–Ukraine war might result in a ceasefire or a largely static front. But whether the war freezes or continues, NATO Allies and Europe will need to continue to support Ukraine through materiel support and security guarantees, to deter further aggression. President Putin has placed Russia on a wartime economy footing that he may be unable or uninterested in departing from. The risk of continuing conflict in the Levant is high. And then there is China's claims in the Pacific, the nine-dash-line and relentless pressure on Taiwan. The global maritime trading system is likely to continue to be pressurized by threats to Taiwan, the Red Sea, the Black Sea, and threats to critical underwater or offshore infrastructure. For the next several years, NATO and the West will likely find themselves in a world with at least three major conflicts either in full swing, near the precipice or in a very uneasy ceasefire. It is not a peacetime environment.

Concrete objectives for the near to mid-term future that a maritime strategy must address include therefore the following: a continuing need to deter Russia from direct aggression against NATO allies leveraging the specific, distinctive contribution sea power makes; continue to support Ukraine; counter China's threats to Taiwan; freedom of navigation and the rules-based order in the Pacific; and preserve the global maritime trading system from challenges to freedom of navigation, including means to counter attacks on critical infrastructure on land and sea, as well as in space and cyberspace.

The ways of a new maritime strategy follow the ends. These have been changing, arguably since 2014, and current NATO and major national defence policies largely reflect acceptance of the primary roles and mission now required. In a parallel to the Cold War, the containment of Russia is based on maintenance of a nuclear strategic balance and conventional superiority to deter aggression against the NATO Alliance. In the maritime domain, this requires a return to the primacy of sea control in some areas, and sea-denial in others. Protection of SLOCs, the capability to embargo or blockade an adversary's trade, and to defend freedom of navigation for one's own return as cardinal points in a new maritime strategy. The need for forward presence continues but not as an adjunct to crisis management operations against non-peer competitors, but rather as a critical deterrent enabler of sea-based carrier striking power and anti-submarine warfare capabilities for deterrence and collective defence. A similar situation exists in the Pacific. This is a marked change from the maritime security and maritime interdiction operations focus of the post-Cold War era (Bergeron, 2024a).

Tying means to geopolitics, strategy is also about identifying the weakness or vulnerability of both the adversary and ourselves, and leveraging our strengths to offset our weaknesses. A maritime strategy truly comes to the fore when a nation or alliance possesses maritime advantage, and further, can use it to offset weaknesses elsewhere. The exemplar of this approach was the 1980s maritime strategy under US Secretary of the Navy John Lehman, which pressed on the Soviet nuclear bastion with naval power to counter the perceived land forces advantages of the Soviet Union on the Fulda Gap and land borders (Lehman, 2018).

So where are the adversary vulnerabilities and alliance maritime strengths that NATO or Western nations might exploit? Framing that question also frames the locations and circumstances in which naval forces are likely to fight. What adversary assumptions might sea power destabilize? For Russia, these arguably include the assumption that the Russian people (or enough of them) can be kept onboard with the Ukraine war amid partial isolation and sanctions, to win the argument that "the war is worth it." A second assumption is that they possess sufficient wealth, reserves, and support from partners such as China and Iran to weather those sanctions and continue to prosecute the war until Western interest fades or politics change. More generally in the competition with NATO, the assumption that any conflict can be geographically contained, kept short (Ukraine notwithstanding) and that NATO is too complex, clumsy, and slow to present an effective challenge to a Russian offensive, probably under a hybrid guise, in the vital short term.

The specific nature of deterrence strategy via Russia, therefore, should leverage naval advantage to threaten outsized destruction of the Russian fleet, that will take a generation to build back, the use of sea power to enforce a radical strengthening of sanctions or blockade in wartime, the capacity to protect transatlantic SLOC, to signal readiness for a prolonged conflict, and to demonstrate the capacity to threaten horizontal escalation.

But there are other challenges from nuclear peer-competitors that may not be deterrable, as they usually exist in the hybrid space, below the threshold of escalation, such as cyber-attack, disinformation, and threats to CUI (Bergeron, 2019). A strategy to counter these challenges is required, that is distinct from both Cold War and post-Cold War strategic options. Finally, the actions of proxies such as the

Houthis do actually parallel the non-peer adversary crisis response operations of the post-Cold War era, and call for multi-domain and all-of-government action to counter, disrupt, and preserve freedom of navigation.

Second RMA and the Transformation of Naval Warfare

Turning to the question of 2RMA, an effective maritime strategy needs to address emerging technology challenges relating to platforms, doctrine, tactics, and budgets. These will impact the future shape of warfare, such as the feasibility of expeditionary operations given new standoff capabilities of relatively modest powers, but also feed back into wider, more political questions such as relations with Russia, China, and Iran, NATO–EU relations, and the division of labour between the US and Europe within NATO. Key issues to be addressed include the balance to be struck between investing in readiness of the legacy fleet or building a new, 2RMA-enabled one; how to retrofit new systems onto old platforms; how to utilize AI in warfare; how new technologies might change deterrence theory and practice; whether emerging systems are rendering flagship legacy platforms like the nuclear aircraft carrier obsolete or not; the future of human-piloted fighter aircraft; the potential future transparency of the oceans; and indeed the environmental costs of 2RMA itself, especially given the mammoth computing power and environmental cost of AI.

The trajectory of the Russia–Ukraine war identifies some important lessons for strategy. Ukrainian drones proved remarkably successful in attacking RFN ships and shore installations in the first half of 2024. Russia responded by the use of nets, maritime patrol aircraft to spot USVs, and attack helicopters to destroy them (Sutton, 2023). Ukraine responded by fixing anti-aircraft missiles onto its USVs, essentially converting them into long-range maritime strike platforms. These have been used successfully against Russian helicopters and against shore targets (Kirichenko, 2025). The point is that these innovations were all accomplished in the space of a year. Competing in this revolution requires incredibly rapid innovation in design and deployment.

Allied navies have been operating autonomous systems for decades. Now they are confronted with a problem of adversary use and rapid modification of inexpensive versions providing both mass and accuracy in littoral operating areas, with potential drone swarms and land-based anti-ship missiles or artillery threatening to overpower the defences of major warships and strike groups, or using the cover of the undersea to attack. These threats raise the costs of littoral power projection.

Navies can respond in several ways, and 2024 saw instances of most of them. The first is to keep calm and carry on. Continue to employ legacy air defence and strike capabilities at high cost and increased risk. This may require avoiding some naval and traditional amphibious operations in the littorals and using other means to defend vital interests there.

A second approach utilizes other existing kinetic systems to contest the littorals at comparable cost. In March 2024, the Italian Frigate CAIO OLIVIO made a conscious choice to engage an incoming drone with its 76mm gun system, which was successful. The Greek Frigate PSARA did the same in July 2024 with its 127mm cannon (Newdick, 2024). The use of gunnery is much less costly and preserves missile stocks but may incur increased risk due to the closer range required for engagement. In February 2025, the US Navy reported test success in using its HELIOS laser to destroy an incoming drone (Ceder, 2025). The different approaches of gunnery or lasers illustrate the fundamental strategic dilemma when faced with asymmetric threats – to aim for countermeasures at the low end of cost and sometimes sophistication, or to escalate innovation and likely expense.

A third approach, perhaps the most novel, was also reportedly employed by PSARA in that encounter, the use of non-kinetic electronic warfare (EW) systems to jam drone communications and navigation (Newdick, 2024). Multiple EW attacks on drones were reportedly conducted by USS MASON and USS GRAVELEY as part of Operation Prosperity Guardian in 2024 (Epstein, 2025).

The final option is to respond in kind. This includes the creation of anti-drone drones, the Torpedo-Boat Destroyers of the 21st century. But a more powerful incorporation of autonomous systems can be found in the new UCAV aircraft carriers developed by Turkey, China, and to a degree Iran (Iddon, 2024). TCG ANADOLU is the world's first carrier designed for an autonomous air wing, the BAYRAKTAR TB3 with a 1000 mile range and a full day's endurance (*Daily Sabah*, 2023). China is building its Type 076 Yulan Class Amphibious Assault Ship to carry an GJ-11 or similar UCAVs. These developments showcase another aspect of 2RMA, the ability it provides for naval powers to expand their maritime strike capabilities and global reach.

There is one large caveat to the new era of autonomy, however, short of overt, great-power conflict. To date, the use of offensive drones and missiles has occurred in contests between non-nuclear powers, or between a nuclear and a non-nuclear power, or militia. Between the nuclear states, drones have been downed by peopled aircraft but an attack on people has not followed an attack on autonomous platforms. In the deterrence equilibrium and its escalation threshold, it appears that autonomous weapons currently exist in a strategic silo (as do the cyber and possibly space domains) and "what happens in Vegas stays in Vegas" (Bergeron, 2022, 11–14). A maritime strategy that incorporates autonomy also needs to address their impact on deterrence when it is perhaps easier to deploy, or destroy, an autonomous weapon than political leaders would have risked with their own or adversary lives at stake (Roberts, 2021).

Technology has not yet seriously impacted the deep ocean nuclear ASW contest, but the basic building blocks are in place: Drug runners can transport tons of cocaine more than 1500 km in autonomous undersea drones. The credibility of a serious 2RMA anti-submarine capability, and the development costs of countering it, is one of the great decisions for a new maritime strategy. The coming transparency of the seas has been falsely prophesied for decades, but it might finally come to pass in this generation. Nations and alliances are already likely inside the decision window to decide if that threat is credible, or when it might be, and develop either new submarine or ASW capabilities to counter the threat or re-think the fundamentals of Western nuclear deterrence policy with its heavy reliance on the sea-based deterrent.

Other innovations in emerging disruptive technologies may also impact the revolution in autonomous capabilities. As with other forms of warfare, at least among the leading powers, advanced autonomous warfare is likely to involve complex, interconnected networks that put a heavy reliance on computing, data, and space assets. The risk of cyber interference or disruption of satellite-based location services presents a challenge to navigation and especially to forms of precision strike. The result is likely to be increased systemic vulnerability, increased costs, and lowered confidence. These effects will impact all parties, and it would be surprising if the US and its Allies were not at the leading edge of this technology. But Western militaries are more reliant on complex networks than their Russian or Houthi counterparts. For maritime strategy, another old dilemma presents itself: the balance between investing in countering threats to the global network, or in expanding capacity for isolated, autonomous action.

Key Enablers of a New Maritime Strategy

The purpose of this chapter was not to offer concrete prescriptions for a new maritime strategy but rather to frame the problem. National circumstances differ greatly, and there is no need for identical solutions, so long as they are interoperable ones. That said, there are a set of cross-cutting key enablers required to deliver whichever choices are made by national or Alliance leaders in responding to the 2RMA and other contemporary challenges.

The first of these is national economic health and the ability to focus a sufficient amount of that on a comprehensive defence strategy. The turning away from global free trade since the mid-2010s has put at risk the benefits in aggregate wealth that free markets are good at delivering. This can be ameliorated through intensified trade and commerce among the like-minded liberal-democratic nations of the West. Facing such concerted opposition from the autocracies, trade wars are a luxury the West can ill afford.

A second key factor in a new maritime strategy is the speed of innovation twinned with the substantial increase in platforms that require years in the building. Although there is some low-hanging fruit in developing quick and mobile autonomous systems, the requirement for deep water sea control and massive power projection does not go away. What is needed is support for a rational defence industrial enterprise, in both the US and Europe. Annual budget cycles, Congressional continuing resolutions and short-term thinking and appropriations cannot give major industries the planning assurance they need to undertake multi-year, if not decades-long, investments in technology and platforms. But equally important is a vibrant entrepreneurial culture in research and development, defence innovation, and acquisition. Highly concentrated markets are seldom competitive. To balance the two, incentives for greater small and medium-sized firm (SME) inclusion in defence innovation, both within and outside the prime contractor model, will be important in re-establishing a robust, competitive defence market. Consideration might also be had to waiving the exemption of the defence industry, both small and large, from large parts of antitrust or competition law in the US and Europe. If innovation is as vital as some suggest, that is a powerful way to harness market forces in its service while preventing abuse.

Sea power is a national project. It is also a grand geopolitical, economic, and ethical project of global dimensions. Successful sea powers have enjoyed cross-party and elite support, over considerable periods of time – matched by high defence budgets, strong government support for logistical infrastructure, R&D, and industry large and small. There is also an essential cultural dimension: sea power needs to run in the blood of the nation (or an alliance) – reflected in media, culture, and imagination – with incentives and rewards for young people to seek a maritime career as both self-actualization and as a wise career move.

Finally, and to that end, education of key stakeholders is essential: a common maritime strategic language is required, shared by naval officers, government and Alliance officials, defence industry managers, and Congressional or Parliamentary aides and special advisors – a language that can connect the ends of policy to the ways and means of a 2RMA era. The Cold War generations possessed this, a result of military service, careers in industry, and a lifetime steeped in the importance of sea power in the 20th century.

That importance needs rekindling at a time of serious international crisis. The project of a new maritime strategy that makes a decisive break with the post-Cold War era, embraces both legacy platforms and emerging 2RMA challenges in creative ways, and frames that relationship in a considered geostrategic theory of victory, can help galvanize government and inspire the public imagination to reassert the Western mastery of sea power for the middle of the 21st century.

References

Berman, Noah. 2024. "How Houthi Attacks in the Red Sea Threaten Global Shipping." *In Brief*. January 12. Council on Foreign Relations. www.cfr.org/in-brief/how-houthi-attacks-red-sea-threaten-global-shipping

Bergeron, James Henry. 2019. "Deterrence and its Maritime Dimension." *The Naval Review* 107 (2): 134–145.

Bergeron, James Henry. 2022. "Dilemmas of Deterrence in an Era of Emerging Destructive Technologies." *Cutting the Bow Wave*. Summer. Combined Joint Operations at Sea Center of Excellence.

Bergeron, James Henry. 2024a. "From Maritime Security to Sea Power: NATO's Paradigm Shift." *ISPI Commentary*. June 11. www.ispionline.it/en/publication/from-maritime-security-to-sea-power-natos-paradigm-shift-176619

Bergeron, James Henry. 2024b. "NATO Maritime Security Operations in the Mediterranean." In *Routledge Handbook of NATO*, edited by John Andreas Olsen. Routledge.

Biddle, Stephen. 1996. "Victory Misunderstood: What the Gulf War Tells Us about the Future of Conflict." *International Security* 21 (2). www.comw.org/rma/fulltext/victory.html

Bogdavov, Constantin and Kramnik, Ila. 2018. *The Russian Navy in the 21st Century*. Center for Naval Analyses. www.cna.org/reports/2018/10/IOP-2018-U-018268-Final.pdf

Ceder, Riley. 2025. "US Navy Hits Drone with HELIOS Laser in Successful Test." *NavyTimes*. February 4. www.navytimes.com/news/your-navy/2025/02/04/us-navy-hits-drone-with-helios-laser-in-successful-test/

Chivvis, Christopher S. and Keating, Jack. 2024. "Cooperation Between China, Iran, North Korea, and Russia: Current and Potential Future Threats to America." *Carnegie Paper*. October 8. https://carnegieendowment.org/research/2024/10/cooperation-between-china-iran-north-korea-and-russia-current-and-potential-future-threats-to-america?lang=en

Cohn, Carolyn. 2024. "Geopolitical Strife Could Cost Global Economy $14.5 Trln Over 5 Years – Lloyd's of London." *Reuters*. October 9. www.reuters.com/markets/geopolitical-strife-could-cost-global-economy-145-trln-over-5-years-lloyds-2024-10-09/

Daily Sabah. 2023. "Türkiye Commissions its Largest Warship, World's 1st Drone Carrier." April 10. www.dailysabah.com/business/defense/turkiye-commissions-its-largest-warship-worlds-1st-drone-carrier

Department of the Navy. 1992. "From the Sea: Preparing the Naval Service for the 21st Century." https://apps.dtic.mil/sti/tr/pdf/ADA338570.pdf

Department of the Navy. 1994. "Forward … From the Sea." https://apps.dtic.mil/sti/pdfs/ADA338561.pdf

Department of the Navy. 2007. "A Cooperative Strategy for 21st Century Sea Power." https://www.govinfo.gov/content/pkg/GOVPUB-D214-PURL-gpo10908/pdf/GOVPUB-D214-PURL-gpo10908.pdf

Dombrowski, Peter and Ross, Andrew L. 2008. "The Revolution in Military Affairs, Transformation and the Defense Industry." *Security Challenges* 4 (4): 13–38.

Epstein, Jake. 2025. "US Destroyers in the Red Sea Conflict Defeated Enemy Weapons Without Firing a Shot, Changing the Way Warships Fight." *Business Insider*. February 6. www.businessinsider.com/us-warships-defeated-drones-without-shooting-changing-how-they-fight-2025-2

Franke, Ulrike. 2025. "Drones in Ukraine: Four lessons for the West." *European Power*. January 10. European Council on Foreign Relations. https://ecfr.eu/article/drones-in-ukraine-four-lessons-for-the-west/

Grady, John. 2024. "Finland Seizes Russian Oil Tanker After Suspected Undersea Fiber-Optic Cable Sabotage." *USNI News*. December 27. https://news.usni.org/2024/12/27/finland-seizes-russian-oil-tanker-after-suspected-undersea-fiber-optic-cable-sabotage

HM Government. 2010. *Securing Britain in an Age of Uncertainty: The Strategic Defence and Security Review*. https://assets.publishing.service.gov.uk/government/uploads/system/uploads/attachment_data/file/62482/strategic-defence-security-review.pdf

House of Commons. 1990. *Defence (Options for Change)*. Parliamentary Debates (Hansard). July 25, col. 468–486.

Iddon, Paul. 2024. "Iran and Turkey Are Betting on Drone Aircraft Carriers to Project Power." *Business Insider*. September 7. www.businessinsider.com/iran-turkey-drone-carrier-ships-project-power-2024-9

Kaushal, Sidharth. 2023. "Stalking the Seabed: How Russia Targets Critical Undersea Infrastructure." *RUSI Commentary*. May 25. https://www.rusi.org/explore-our-research/publications/commentary/stalking-seabed-how-russia-targets-critical-undersea-infrastructure

Kaushal, Sidharth; Byrne, James; Byrne, Joe; and Somerville, Gary. 2021. "The Yasen-M and the Future of Russian Submarine Forces." *RUSI Defence Systems* 23. https://www.rusi.org/explore-our-research/publications/rusi-defence-systems/yasen-m-and-future-russian-submarine-forces

Kauranen, Anne; Lehto, Essi; and Rinke, Andreas. 2025. "NATO to Deploy Ships, Aircraft in Baltic Sea after Cable Breaches." *Reuters*. January 14. www.reuters.com/world/europe/baltic-sea-nations-seek-limit-further-incidents-after-cable-breaches-2025-01-14/

Kirichenko, David. 2025. "Ukraine's Marauding Sea Drones Bewilder Russia." *Europe's Edge*. January 30. CEPA.

Lambert, Andrew. 2018. *Seapower States: Maritime Culture, Continental Empires and the Conflict That Made the Modern World*. Yale University Press.

Lawford, Melissa. 2024. "Red Sea Attacks Trigger Fresh Inflation Fears as Shipping Plunges." *The Telegraph*. July 5.

Lehman, John F. 2018. *Oceans Ventured: Winning the Cold War at Sea*. W. W. Norton & Co.

Mahan, Alfred Thayer. 1890. *The Influence of Sea Power Upon History, 1660–1783*. Little Brown.

Meade, Richard. 2024. "Houthi Threat to Shipping Growing Thanks to 'Unprecedented' Network of Support." *Lloyd's List*. November 4. www.lloydslist.com/LL1151228/Houthi-threat-to-shipping-growing-thanks-to-unprecedented-network-of-support

NATO. 2011. *Alliance Maritime Strategy*. www.nato.int/cps/da/natohq/official_texts_75615.htm

Nevola, Luca and d'Hauthuille, Valentin. 2024. "Six Houthi Drone Warfare Strategies: How Innovation Is Shifting the Regional Balance of Power." *ACLED*. August 6. https://acleddata.com/2024/08/06/six-houthi-drone-warfare-strategies-how-innovation-is-shifting-the-regional-balance-of-power/

Newdick, Thomas. 2024. "Greek Warship Guns-Down Houthi Drone In New Video." *The Warzone*. July 9. www.twz.com/sea/greek-warship-guns-down-houthi-drone-in-new-video

O'Connell, Richard. 1993. *Sacred Vessels: The Cult of the Battleship And the Rise of the U.S. Navy*. Oxford University Press.

Padfield, Peter. 1999. *Maritime Supremacy & the Opening of the Western Mind: Naval Campaigns That Shaped the Modern World*. Overlook Books.

Roberts, Brad. 2021. *Emerging and Disruptive Technologies, Multi-domain Complexity, and Strategic Stability: A Review and Assessment of the Literature*. Centre for Global Security Research, Lawrence Livermore National Laboratory. https://cgsr.llnl.gov/content/assets/docs/EDT_ST2_BHR_2021.3.16.pdf

Shuster, Simon. 2024. "The Drone Wars: How Ukraine Beat Russia in the Battle of the Black Sea." *TIME*. August 26. https://time.com/7013531/sea-drones-how-ukraine-beat-russia-in-the-black-sea/

Slayton, Nicholas. 2024. "Cheap Houthi Drones Are Draining the Pentagon's Coffers." *New Lines Magazine*. July 29. https://newlinesmag.com/argument/cheap-houthi-drones-are-draining-the-pentagons-coffers/

Sommerville, Quentin. 2024. "Ukraine Thrown Into War's Bleak Future as Drones Open New Battlefront." *BBC News*. July 24. www.bbc.co.uk/news/articles/cne4vl9gy2wo

Sutton, H I. 2021. "Russia's Growing Secret Submarine Fleet Key to Moscow's Undersea Future." *USNI News*. November 30. https://news.usni.org/2021/11/30/russia-growing-secret-submarine-fleet-key-to-moscows-undersea-future

Sutton, H I. 2023. "Russia Forced to Adapt to Ukraine's Maritime Drone Warfare in Black Sea." *Naval News*. December 2. www.navalnews.com/naval-news/2023/12/russia-forced-to-adapt-to-ukraines-maritime-drone-warfare-in-black-sea/

Sutton, H I. 2024. "Uncrewed Platforms Have Been Critical to Ukraine's Success in the Black Sea." *RUSI Commentary*. August 20. www.rusi.org/explore-our-research/publications/commentary/uncrewed-platforms-have-been-critical-ukraines-success-black-sea

United Kingdom Ministry of Defence. 1998. *Strategic Defence Review White Paper*. https://researchbriefings.files.parliament.uk/documents/RP98-91/RP98-91.pdf

United States Department of Defence. 2002. "Defense.gov News Transcript: DoD News Briefing – Secretary Rumsfeld and Gen. Myers." February 12.

United States Navy. 2024. *Chief of Naval Operations Navigation Plan for America's Warfighting Navy 2024*. https://news.usni.org/2024/09/18/cno-franchettis-new-navy-navigation-plan

9 Preparing Civilian Infrastructure for Potential Cyber and Hybrid Attacks

Konstantinos Tsetsos

Introduction

In an era marked by increasing complexity and interconnectivity, hybrid and cyber threats pose a significant challenge to modern societies. Combining tactics such as disinformation campaigns, strategic corruption, economic coercion, and cyberattacks, these threats aim to destabilize critical infrastructures (CI) and social cohesion while operating in the ambiguous "grey zone" between peace and conflict. This chapter explores the multifaceted nature of hybrid threats, their impact on CIs, and the strategies needed to counteract them effectively. With a focus on resilience, cybersecurity, international collaboration, and proactive measures, this chapter offers a comprehensive look into how nations and organizations can enhance their resilience vis-à-vis kinetic and non-kinetic hybrid threats.

Defining Hybrid and Cyber Threats

Hybrid threats combine various methods and tactics to achieve political, economic, or military goals such as destabilizing states or societies. The main characteristic is the concealment of authorship and the use of a "grey area" between war and peace. They often operate below the threshold of open conflict and thus avoid clear international reactions (Tsetsos, 2021). Hybrid threats represent the most recent strategic development in military affairs. They break with traditional understandings of war and warfare, intertwine old and new approaches, and challenge existing defense and resilience approaches.

The first three generations of warfare focused on conventional warfare between states. First-generation warfare (formation warfare) prevailed from antiquity to the 19th century and was characterized by line and column formations of uniformed heavy infantry. The goal was to physically shove an opponent off the field of battle through the application of force – winning multiple battles or one decisive engagement constituted victory. Second-generation warfare (firepower warfare) was dominated by greater accuracy and firepower of long-range weapons, rail transport, and motorization as well as increasing industrialization of the war economy between 1850 and 1930. This increase in firepower led to arms races, and attritional and trench

DOI: 10.4324/9781003520160-9

warfare, while the industrialization of conflict also emphasized the economic capacity to wage war. The third generation of warfare (maneuver warfare) focused on combined arms operations based on tactics of speed and surprise to overcome potential deadlocks. The aim was to bypass the enemy's lines and collapse their forces from the rear in a swift manner. The focus in these first three generations of warfare was on the physical destruction of enemy armed forces (Tsetsos, 2023).

Fourth-generation warfare (decentralized use of force) is aimed at undermining the psychological ability of an adversary to conduct warfare by using public pressure to force the hands of political decision-makers (Hammes, 2006). Insurgents primarily use indirect warfare against the state to cause military casualties, especially in democratic states with a high level of casualty aversion. The civilian population, public opinion, and decision-makers thus become the primary strategic focus. Fifth-generation warfare (non-kinetic attacks) is dominated primarily by social engineering, the spreading of false information, cyber-attacks, and the use of artificial intelligence (AI) and autonomous systems (Abbott, 2010). Here, too, the aim is to influence the will of the public and their decision-makers by using non-kinetic means and technological innovations. These non-kinetic attacks focus on the maximization of the attribution problem to mask their origin and maintain plausible deniability through obfuscation.

All of these generations of warfare are ideal-type forms of war which are not mutually exclusive and may be applied simultaneously. States today thus face both conventional and hybrid threats which together constitute complex and dynamic security challenges. The non-traditional means used in this context range from strategic corruption, electoral interference, and economic weakening to the planned dissemination of propaganda, cyber activities, and espionage. Cyber-attacks that undermine infrastructure, financial systems, or government institutions and disinformation campaigns that establish counter narratives are on the rise. What is more, attacks by state and non-state actors operating in a hybrid manner expand into other areas of society. In addition to conventional and asymmetric kinetic attacks like terror, and sabotage, CI will be increasingly faced with hybrid threats such as cyberattacks, disinformation campaigns, economic pressure, and subversive actions. Cyber threats are a central component of hybrid threats using weak points in digital networks to attack CI. These methods target the vulnerability of a society, especially CI and social cohesion, and make it difficult to identify clear aggressors (Tsetsos, 2023). As a result, hybrid warfare goes beyond a pure threat and refers to a strategic escalation in which military and non-military means are combined to achieve a goal. Examples include election manipulation, strategic corruption, propaganda, and cyber-attacks, as was seen in the Ukraine conflict. Russia and China are seen as pioneers in this area, as they specifically use hybrid approaches to destabilize Western states (Hodges, 2021).

Measures to Counter Hybrid Threats

In order to effectively counter cyber and hybrid threats, a combination of preventive, reactive, and strategic measures that involve both state and societal actors is warranted. The most important recommendations for action and measures can be summarized as follows (Tsetsos, 2020):

- Promoting Resilience: Strengthen the ability to absorb attacks with robust infrastructure, coordinated state-wide approaches, and comprehensive crisis response training.
- Deterrence by Resilience: Increase adaptability and minimize vulnerabilities to deter attackers through system resilience, and combat disinformation using detection technologies, media/digital literacy, and public education.
- Enhancing Cybersecurity: Implement advanced technical defenses (e.g., firewalls, multi-factor authentication), establish security operations centers (SOC), and focus on securing CIs like energy and communication networks.
- International Collaboration: Foster close NATO/EU cooperation, share resources, and standardize cybersecurity measures and incident reporting.
- Awareness and Training: Raise public and corporate awareness of hybrid threats and conduct regular practical and virtual crisis exercises to enhance response capabilities.

In summary, hybrid threats exploit the weaknesses of modern societies, in particular their dependence on digital and globalized networks. Studies emphasize that an effective defense requires not only technical resilience, but also social resilience. This includes strengthening CI and promoting social cohesion and international cooperation in dealing with cyber and hybrid threats (Tsetsos, 2020).

Definition and General Discourse on CI

CI is defined as "buildings, facilities, systems or networks essential for maintaining the vital functions of a society, and the health, safety, security and economic and social well-being of the community, whose cessation or destruction would have a significant impact" (Curt & Tacnet, 2018, p. 2441). It is also referred to as the economy's nerve system enabling societal functioning while also driving economic growth and fostering social development (Yusta et al., 2011).

The sectors that a country defines as critical may vary. For example, among the CI sectors of the U.S. are chemical, communications, critical manufacturing, dams, defense, emergency services, energy, financial services, food and agriculture, government facilities, healthcare, Information Technology, Nuclear Reactors, Waste, Transportation Systems, and Water Systems (U.S. Department of Homeland Security 2024). These sectors are interconnected and partly depend on each other implying that one CI may rely on the services of another to operate effectively (Alcaraz & Zeadally, 2015).

Early discussions about threats to CI primarily centered on physical threats, including terrorism, natural disasters, and other destructive forces. More recently, non-physical threats, such as cyberattacks, and political challenges emerged as significant concerns in the literature (Biskupovic, 2021). The number of cyberattacks on CI has risen sharply in recent years. A report by Forescout-Vendere Research

(2024) indicates that from January 2023 to January 2024, global CI experienced over 420 million cyberattacks, which equals more than 13 attacks per second and an increase of 30% compared to the previous year. The following sections are consequently dedicated to this topic.

The Increasing Risk of Cyberattacks on CI

Advancements in technology over recent decades have made CI increasingly reliant on digital operating systems and network infrastructure (Osei-Kyei et al., 2021). This growing dependence led to an integration of formerly isolated CI operating systems into a system framework, where functionality and productivity depend on public networks like the internet. That global interconnectivity has made CI a prime target for cyberattacks, with scholars highlighting the rising frequency of such incidents (Gunduz & Das, 2020). This is also due to the decreased time needed to plan an attack and the ease with which advanced technologies can be leveraged to target CI (Han et al., 2019).

Attacks on the technological systems underpinning CI can have severe consequences. A disrupted cyber infrastructure could lead to significant impacts on the performance, reliability, and security of CI due to their strong reliance on interconnected systems (Alcaraz & Zeadally, 2015). Moreover, the interdependence of CI can result in cascading effects (Alcaraz & Lopez, 2012). For example, in 2016, presumably Russian hackers carried out a cyber-attack on the Ukrainian power facility Prykarpattya Oblenergo in December causing a blackout that left nearly half the population – approximately 700,000 people – of Ukraine's Ivano-Frankivsk region without electricity affecting further CI including healthcare and communications (Weinberg, 2021). Scenario planning has attempted to address such cascading effects by cross-impact analysis. For example, a hypothetical attack on a major European port such as Antwerp or Hamburg that results in its defunction for a week is anticipated to cause detrimental cascading effects that ripple through the entire European production and economic supply chain. While damage and loss estimates are impossible to predict, it is a common opinion amongst experts that the tally of such event could cause an economic cost/loss of dozens of billions of Euros. Thus, safeguarding CIs from these attacks remains a daunting task due to the overwhelming volume of security events that require analysis and response (Han et al., 2019). This is why it is essential to better understand cyber-attacks, their impact, and the offenders' motivation in order to design countermeasures.

Types and Impacts of Cyber-Attacks

The classification of cyber threats by Darem et al. (2023) provides a framework to understand the complexities of cyberattacks, which is applicable across CI sectors (Ige et al., 2024). It includes the *impact*, *nature*, and *character* of cyber threats (Darem et al., 2023). *Impacts* range from regional to global economic destabilization (Sheehan et al., 2021), financial losses (Mukhopadhyay et al., 2019), reputational damage (Wang et al., 2020), operational disruption (Kaffenberger &

Kopp, 2019), and increased costs or compliance penalties (Sheehan et al., 2021). Cyberattacks can also result in intellectual property theft and identity breaches (Stanikzai & Shah, 2021). Cyber threats are further categorized by their *nature* and *character* (Darem et al., 2023). The *nature* includes the rapid development, covertness as cyberattacks are hidden, and adversarial threats from state-sponsored actors, crime syndicates, and hacktivists (Vedral, 2021). Analyzing identified threat actors in 2023, the majority of them were either criminals (47%) or state-sponsored actors (46%) (Forescout-Vendere Research, 2024). Threat *characteristics* include insider involvement, that is, manipulation or recruitment of employees (Canadian Centre for Cyber Security, 2023), persistence, and adaptability of hacker tactics (Stanikzai & Shah, 2021), and multi-vector attacks like malware and phishing (Mattioli et al., 2023).

Finally, the threat is characterized by the *motivation* of the attack ranging from financial to political goals (Doerr et al., 2022), with attacks often targeting CI institutions (Sood & Enbody, 2014). In 2023, CI organizations in 163 countries were targeted, with the US (28%) and UK (14%) being top targets, followed by Germany with 12.8% (Forescout-Vendere Research, 2024). The most active threat actors were based in China (25.8%), Russia (14.7%), and Iran (7.5%). The CI most frequently targeted by these actors were Government, Financial Services, Media and Entertainment, Technology, Education, Telecommunication, Healthcare, and Energy (Forescout-Vendere Research, 2024). Responding to cyberattacks is complicated by a lack of skilled professionals and inconsistencies in organizational security priorities that may create potential entry points for attackers (Han et al., 2019; Zimba et al., 2018). Recognizing common threats and their characteristics is thus essential for designing effective countermeasures.

Literature Review on Countermeasures

The literature emphasizes both technical and non-technical measures – legal, regulatory, and organizational – in mitigating cyber risks to CI (Darem et al., 2023). Technical measures include encryption, a key tool to prevent unauthorized data access (Diro et al., 2020), alongside multi-factor authentication, network segmentation (Basta et al., 2022), firewalls, and intrusion prevention systems (IPS) (Zahoor et al., 2016). Regular updates, endpoint protection, security information and event management (SIEM), data loss prevention (DLP), anomaly detection, and deception technologies further bolster security (González-Granadillo et al., 2021; Kechagias et al., 2022; Diro et al., 2020).

Legal and regulatory measures like general data protection regulation (GDPR) and ISO 27001 establish essential cybersecurity standards (Sudarwanto & Kharisma, 2022; Haruna et al., 2022). Breach notification policies, cyber insurance, and international collaboration aid in incident response and reducing risks (Kosseff, 2017; Matejka et al., 2021). *Organizational measures* include incident response planning, risk assessment, regular audits, and awareness training (Yohannes et al., 2019; Kennedy, 2016). SOC, chief information security officers, employee screening, threat intelligence sharing, and disaster recovery plans enhance readiness (Danquah, 2020; Wagner et al., 2019). Physical security like closed-circuit television (CCTV) completes these strategies (Badhwar & Badhwar, 2021). Due to the unique requirements and risks faced by individual organizations, they must tailor these measures to their specific needs, integrating policies, and technical and procedural strategies for robust protection (Darem et al., 2023).

Recommended Courses of Action

More robust states, companies, and societies are better at weathering crises. They tend to recover more quickly and are able to return to their pre-crisis level of functioning. Less robust societies are paralyzed by crises for longer and thus miss out on other political, economic, or social developments, running the risk of trailing behind progress for years. The primary objective of resilience is to ensure the continuity of government and essential public services even in a state of emergency. Resilience is especially improved if, in addition to government preparations, resources of the civilian sector are used to support state tasks. A number of future-oriented measures at the national and European level can contribute to strengthen civilian CI operators in particular and the national resilience overall.

Recommendation 1: Strengthening National Information Exchange

Disaster and crisis management is subject to the principle of subsidiarity and as such is mostly the responsibility of the individual sub-state entities. A common operational picture on the national level that combines civil protection, crisis management, civilian and military stakeholders, and national defense through a joint information and situation center is a prerequisite for adequate resilience. Information exchange requires institutionalized and voluntary cooperation. Data that could be shared with command-and-control information (C2I) systems of the civilian and military response forces involved are usually only sporadically available. Because there are a number of different systems in use, mutually agreed standards for information exchange among civilian authorities and between civilian and military response forces are also lacking. To overcome such shortcomings, a number of initiatives seem appropriate:

- expanding cooperation and information exchange between government authorities and the civilian sector, including non-governmental organizations (NGOs)
- defining technical standards and harmonizing different monitoring and C2I systems on a national level
- establishing permanent monitoring based on a joint situation picture
- expanding civil–military cooperation on the management of hybrid risks and natural disasters

Recommendation 2: Strengthening International Information Exchange

The deficits that affect national information exchange also exist in the international context. While there are established NATO standards in the area of simulation-based military training, they are primarily geared towards military operations and are not explicitly intended for crisis management, operation to counter hybrid attacks, or resilience strengthening. There are a number of ways in which international cooperation could be improved:

- creating European standards for training civilian and military personnel and establishing a reserve of skilled civilian volunteers in the EU who can be called up in the event of a crisis as a result of an attack on CI
- creating European standards for EU-wide information exchange before, during and after crises
- expanding cross-border civil-military cooperation in crises
- compiling a common operational picture for all of Europe to increase resilience
- Europeanizing CI protection
- defining and cataloguing national and European CI
- establishing EU-wide permanent situation monitoring and situational awareness with regard to CI
- creating cross-border emergency and contingency plans, joint safeguard measures and transnational redundancies
- building a European smart grid to safeguard power supply
- establishing minimum standards for cyber resilience for central state tasks

Recommendation 3: Establishing Early Recognition of Crises to Support Decision-Making Processes

There currently are a variety of ways to measure and predict crises and potential vulnerabilities. What is lacking, however, is a process for translating the results into political action. Ways to improve this situation include:

- making better use of early warnings for decision support
- developing processes for early containment of slowly developing crises

Recommendation 4: Promoting a Resilience-Based Security Culture

Resilience is about permanent transformation, not just maintaining the status quo. The careful handling of political and technical security as well as the adaptation to disruptive emerging technologies is key. Investments in security must not be perceived as a drain. Instead, they should be considered a permanent part of the organizational culture of both government and the private sector and should be exercised and rehearsed in realistic scenarios. The following measures can help establish a resilience-based security culture:

- promoting a resilience-based security culture across all areas of society through national, European, and NATO guidelines
- incentivizing prevention measures to establish them as something other than bothersome bureaucratic requirements, for example, by reducing the tax burden on companies that invest in reducing their vulnerability to crises

- increasing awareness for security and safety through crisis and resilience plans
- developing national and EU-wide crisis scenarios and performing regular exercises as stress tests

Conclusion

As the landscape of security threats evolves, the imperative to prepare civilian infrastructure for potential cyber and hybrid attacks becomes increasingly urgent. The convergence of digital dependence, interconnectivity, and sophisticated hybrid threat tactics has made CI a prime target, with cascading effects that can disrupt entire societies. The analysis presented underscores the need for a multifaceted approach combining technical, organizational, and strategic measures.

Hybrid warfare, particularly its fourth- and fifth-generation manifestations, poses significant challenges to the ability of civilian and military actors to prepare for and protect CI. The decentralized and psychological focus of fourth-generation warfare disrupts traditional defense mechanisms by targeting societal cohesion and public confidence, often bypassing physical defenses entirely. Meanwhile, fifth-generation warfare leverages technological advancements, such as AI and cyberattacks, to exploit vulnerabilities in interconnected systems while maintaining plausible deniability. These methods obscure the origin of attacks, complicating attribution and response, and create an environment where proactive defense measures must constantly evolve to address emerging threats. The resulting uncertainty and operational complexity strain both civilian and military preparedness, requiring a seamless integration of technological innovation, policy adaptation, and cross-sector collaboration to safeguard CI effectively.

Resilience remains at the heart of preparedness, requiring not only robust cybersecurity but also coordinated crisis response, public awareness, and international collaboration. Strengthening information exchange, establishing clear standards, and fostering a resilience-based security culture are essential steps to mitigate vulnerabilities and enhance recovery capabilities. Moreover, the integration of civilian and governmental resources, supported by realistic scenario planning and training, can further bolster national and international defenses.

Ultimately, addressing the challenges posed by cyber and hybrid threats demands a proactive and adaptive approach. Governments, industries, and communities must work in tandem to transform security from a reactive measure into a foundational element of societal stability and resilience. By doing so, societies can ensure the continuity of essential services, safeguard their CIs, and remain resilient in the face of emerging threats.

References

Abbott, Daniel. 2010. *The Handbook of Fifth-Generation Warfare*. Nimble Books.

Alcaraz, Cristina, and Javier Lopez. 2012. "Analysis of Requirements for Critical Control Systems." *International Journal of Critical Infrastructure Protection* 5 (3–4): 137–145.

Alcaraz, Cristina, and Sherali Zeadally. 2015. "Critical Infrastructure Protection: Requirements and Challenges for the 21st Century." *International Journal of Critical Infrastructure Protection* 8: 53–66.

Badhwar, Rishi, and Ruchi Badhwar. 2021. "Cybersecurity Lessons from the Breach of Physical Security at US Capitol Building." In *The CISO's Transformation: Security Leadership in a High Threat Landscape*, 125–29. Springer.

Basta, Nektarios, Muhammad Ikram, Mohamed Ali Kaafar, and Andrew Walker. 2022. "Towards a Zero-Trust Micro-Segmentation Network Security Strategy: An Evaluation Framework." In *NOMS 2022-2022 IEEE/IFIP Network Operations and Management Symposium*, 1–7. IEEE.

Biskupovic, Aleksandra. 2021. "Critical Infrastructure Resilience: Findings from a Systematic Review." Master's thesis, University of Waterloo.

Canadian Centre for Cyber Security. 2023. *An Introduction to the Cyber Threat Environment*. www.cyber.gc.ca/en/guidance/introduction-cyber-threat-environment#defn-compromise.

Curt, Christine, and Jean-Marc Tacnet. 2018. "Resilience of Critical Infrastructures: Review and Analysis of Current Approaches: Resilience of Critical Infrastructures." *Risk Analysis* 38 (11): 2441–58. https://doi.org/10.1111/risa.13166.

Danquah, Prince. 2020. "Security Operations Center: A Framework for Automated Triage, Containment and Escalation." *Journal of Information Security* 11 (4): 225–40.

Darem, Abdulbasit A., Asma A. Alhashmi, Tareq M. Alkhaldi, Abdullah M. Alashjaee, Sultan M. Alanazi, and Shouki A. Ebad. 2023. "Cyber Threats Classifications and Countermeasures in Banking and Financial Sector." *IEEE Access* 11: 125138–125158.

Diro, Adane, Hailu Reda, Nadeem Chilamkurti, Akhlaq Mahmood, Nizam Uddin Zaman, and Yong Nam. 2020. "Lightweight Authenticated-Encryption Scheme for Internet of Things Based on Publish-Subscribe Communication." *IEEE Access* 8: 60539–51.

Doerr, Sebastian, Leonardo Gambacorta, Thomas Leach, Benedikt Legros, and David Whyte. 2022. "Cyber Risk in Central Banking." Government Canada, Ottawa, Canada: Bank for International Settlements Financial Stability Institute.

Forescout-Vendere Research. 2024. "2023 Global Threat Roundup." www.forescout.com/resources/research-report_2023-threat-roundup.

González-Granadillo, Graciela, Sara González-Zarzosa, and Rafael Diaz. 2021. "Security Information and Event Management (SIEM): Analysis, Trends, and Usage in Critical Infrastructures." *Sensors* 21 (14): 4759.

Gunduz, Murat Zaim, and Resul Das. 2020. "Cyber-Security on Smart Grid: Threats and Potential Solutions." *Computer Networks* 169: 107094.

Hammes, Thomas X. 2006. "*Sling and the Stone: On War in the 21st Century*." Zenith Press.

Han, Chul Ho, Sang Tae Park, and Seung Jae Lee. 2019. "The Enhanced Security Control Model for Critical Infrastructures with the Blocking Prioritization Process to Cyber Threats in Power System." *International Journal of Critical Infrastructure Protection* 26: 100312.

Haruna, Wasiu, Tunde Aremu, and Yetunde Modupe. 2022. "Defending Against Cybersecurity Threats to the Payments and Banking System." *arXiv preprint* arXiv:2212.12307.

Hodges, Ben. 2021. "Lt-Gen Ben Hodges on the Future of Hybrid Warfare." https://cepa.org/article/lt-gen-ben-hodges-on-the-future-of-hybrid-warfare/.

Ige, Anthony B., Ebenezer Kupa, and Olu Ilori. 2024. "Analyzing Defense Strategies Against Cyber Risks in the Energy Sector: Enhancing the Security of Renewable Energy Sources." *International Journal of Science and Research Archive* 12 (1): 2978–95.

Kaffenberger, Lukas, and Elias Kopp. 2019. *Cyber Risk Scenarios, the Financial System, and Systemic Risk Assessment*. Washington, DC: Carnegie Endowment for International Peace.

Kechagias, Emmanuel P., George Chatzistelios, George A. Papadopoulos, and Panagiotis Apostolou. 2022. "Digital Transformation of the Maritime Industry: A Cybersecurity Systemic Approach." *International Journal of Critical Infrastructure Protection* 37: 100526.

Kennedy, Shawn E. 2016. "The Pathway to Security–Mitigating User Negligence." *Information & Computer Security* 24 (3): 255–64.

Kosseff, Jeff. 2017. "Defining Cybersecurity Law." *Iowa Law Review* 103: 985.

Matejka, Viktor, Jose Soto, and Manuel Franco. 2021. "A Framework for the Definition and Analysis of Cyber Insurance Requirements." Master's project, Zurich, Switzerland: University of Zurich, Communication Systems Group, Department of Informatics.

Mattioli, Riccardo, Andreas Malatras, Edward N. Hunter, Marco G. B. Penso, Daniel Bertram, and Ingo Neubert. 2023. "Identifying Emerging Cyber Security Threats and Challenges for 2030." *European Union Agency for Cybersecurity (ENISA)*, Athens-Heraklion, Greece, Technical Report 64.

Mukhopadhyay, Anirban, Saptarshi Chatterjee, and Kishore Kumar Bagchi. 2019. "Cyber Risk Assessment and Mitigation (CRAM) Framework Using Logit and Probit Models for Cyber Insurance." *Information Systems Frontiers* 21: 997–1018. https://doi.org/10.1007/s10796-017-9808-5.

Osei-Kyei, Robert, Vivian Tam, Michael Ma, and Frank Mashiri. 2021. "Critical Review of the Threats Affecting the Building of Critical Infrastructure Resilience." *International Journal of Disaster Risk Reduction* 60: 102316.

Sheehan, Brian, Fergus Murphy, Alireza Nourani Kia, and Richard Kiely. 2021. "A Quantitative Bow-Tie Cyber Risk Classification and Assessment Framework." *Journal of Risk Research* 24 (12): 1619–38.

Sood, Aditya, and Richard Enbody. 2014. *Targeted Cyber Attacks: Multi-Staged Attacks Driven by Exploits and Malware*. Rockland, MA: Syngress Media.

Stanikzai, Abdul Qayum, and Mirza Aman Shah. 2021. "Evaluation of Cyber Security Threats in Banking Systems." In *2021 IEEE Symposium Series on Computational Intelligence (SSCI)*, 1–4. IEEE.

Sudarwanto, Achmad S., and Daniel B. B. Kharisma. 2022. "Comparative Study of Personal Data Protection Regulations in Indonesia, Hong Kong, and Malaysia." *Journal of Financial Crime* 29 (4): 1443–57.

Tsetsos, Konstantinos. 2020. "Resilience Thinking." Metis Study No. 21. https://metis.unibw.de/assets/pdf/metis-studie21-2020_11-resilienz.pdf.

Tsetsos, Konstantinos. 2021. "New Hybrid Threats." Metis Study No. 26. https://metis.unibw.de/assets/pdf/metis-study26-2021_07-hybrid_threats.pdf.

Tsetsos, Konstantinos. 2023. "Trends and Developments in Hybrid Threats." Metis Study No. 35. https://metis.unibw.de/assets/pdf/metis-study35-2023_06-trends_and_developments_in_hybrid_threats.pdf.

U.S. Department of Homeland Security. 2024. *Critical Infrastructure Sectors*. www.cisa.gov/topics/critical-infrastructure-security-and-resilience/critical-infrastructure-sectors.

Vedral, Borut. 2021. "The Vulnerability of the Financial System to a Systemic Cyberattack." In *Proceedings of the 13th International Conference on Cyber Conflict (CyCon)*, 95–110.

Wagner, Tobias D., Kazi Mahbub, Eduardo Palomar, and Abdelrahman E. Abdallah. 2019. "Cyber Threat Intelligence Sharing: Survey and Research Directions." *Computers & Security* 87: 101589.

Wang, Victor, Huy Nnaji, and Jin Jung. 2020. "Internet Banking in Nigeria: Cyber Security Breaches, Practices, and Capability." *International Journal of Law, Crime and Justice* 62: 100415.

Weinberg, Ariel. 2021. "Analysis of Top 11 Cyber Attacks on Critical Infrastructure." www.firstpoint-mg.com/blog/analysis-of-top-11-cyber-attackson-critical-infrastructure/.

Yohannes, Tsedale, Leila Lessa, and Solomon Negash. 2019. "Information Security Incident Response Management in an Ethiopian Bank: A Gap Analysis." *AMCIS 2019 Proceedings*. 22, Cancun. https://aisel.aisnet.org/amcis2019/adv_info_systems_research/adv_info_systems_research/22/

Yusta, Jose M., Guillermo J. Correa, and Ricardo Lacal-Arántegui. 2011. "Methodologies and Applications for Critical Infrastructure Protection: State-of-the-Art." *Energy Policy* 39 (10): 6100–1119. https://doi.org/10.1016/j.enpol.2011.07.010.

Zahoor, Zulfiqar, Muhammad Ud-din, and Koichiro Sunami. 2016. "Challenges in Privacy and Security in Banking Sector and Related Countermeasures." *International Journal of Computer Applications* 144 (3): 24–35.

Zimba, Armstrong, Zhiqiang Wang, and Hongbin Chen. 2018. "Multi-Stage Crypto Ransomware Attacks: A New Emerging Cyber Threat to Critical Infrastructure and Industrial Control Systems." *ICT Express* 4 (1): 14–18.

10 The Role of Military and Their Training in High-Tech Warfare

Fotios Moustakis

Introduction

We live in a world defined by rapid and unpredictable change, with geopolitics at its core. Once-stable power dynamics are now in flux, and disruptive behaviour is increasingly the norm. Western nations face a complex web of interconnected challenges, while their geopolitical adversaries are seizing the opportunity to exploit this period of uncertainty. Free from the constraints of democratic systems—such as public opinion, political opposition, and fair elections—these adversaries are able to think creatively, set long-term goals, and implement strategies with greater agility. In contrast, the West risks being outpaced and outthought by its competitors, who benefit from centralised decision-making and the ability to act swiftly.

One of the most significant factors exacerbating this imbalance is the rapid advancement of transformative technologies. These technologies, including artificial intelligence (AI), robotics, cyber warfare, and autonomous systems, have the potential to fundamentally reshape military and strategic landscapes. In many ways, they threaten the traditional foundations of Western military supremacy, which has long been grounded in superior firepower, advanced weaponry, and well-trained personnel. The diffusion of cutting-edge technologies to adversaries raises critical questions: How will Western military organisations adapt? Can they maintain their edge in a world where technological innovations evolve faster than existing doctrines and capabilities?

The challenge is not only about technological adoption but also about leadership. The rise of these transformative technologies presents unprecedented leadership dilemmas. As new systems and tools reshape how wars are fought and how strategies are developed, military and organisational leaders will need to adapt their approach to managing both technology and people. Effective leadership in this new era will require more than just technical competence; it will demand vision, agility, and the ability to understand the human dimension of warfighting and organisational dynamics.

Human leadership has a distinct comparative advantage in areas such as motivating troops, and fostering the kind of innovation and talent that technology alone cannot produce. While new technologies will undoubtedly provide critical insights

DOI: 10.4324/9781003520160-10

and efficiencies, they cannot replace the uniquely human skills needed to connect with and inspire people. Transforming these technological advancements into operational successes will require leaders who can effectively communicate the significance of these tools to their teams and align technological capabilities with organisational objectives. The Russian invasion of Ukraine has provided a stark illustration of this dynamic. Despite significant material disadvantages, Ukraine's adaptive leadership and innovative use of technology have allowed it to resist a more powerful adversary effectively. The conflict highlights the importance of empowering leaders at all levels to act decisively, leverage new technologies, and adapt to rapidly shifting conditions on the battlefield.

This chapter argues that, in order to meet these emerging challenges, a shift is required in how leadership is conceptualised and practiced in the military. Conventional transactional leadership models, which focus on short-term exchanges and performance metrics, are insufficient in this context. Instead, a transformational leadership approach is needed—one that prioritises long-term vision, the development of individuals, and the ability to foster innovation and adaptability within organisations. Drawing insights from the military context, this chapter underscores the critical role of leaders who can inspire, empathise, and cultivate talent, especially as they navigate the complexities of emerging technologies.

An Overview of New Technologies in Global Military Strategy

Before examining the role of soldiers in the current era of high-tech warfare, it is important to first assess the role and impact of transformative technologies on global military strategy. The ability to contextualise machine-driven decisions is expected to become a crucial asset for future military leaders, with its effective use likely to shape and define the most suitable military leadership and training styles for the 21st century.

Recent conflicts, particularly the Ukraine war, have underscored the growing importance of technologies like drones and AI. For example, Ukrainian forces have extensively employed drones, including Bayraktar TB2 and domestically produced models, for reconnaissance, targeting, and psychological operations. This use of commercial and military-grade drones has provided Ukrainian forces with critical advantages, enabling them to disrupt Russian supply lines, gather real-time intelligence, and conduct precision strikes.

According to the Oxford Living Dictionary, AI is defined as:

> "The theory and development of computer systems able to perform tasks normally requiring human intelligence, such as visual perception, speech recognition, decision-making, and translation between languages" (Marr, 2018). At the organizational level, AI, is a "cognitive technology" which will "enable organizations to break prevailing trade-offs between speed, cost, and quality," increasing efficiencies and output (Jensen et al., 2020, 526–550). On a strategic level, the application of AI could enable emerging powers like China to displace existing military powers such as the United States.

AI does not refer to just one technology, but rather to a collection of them. Most of these technologies are loaded with latent military potential (Davis and Nacht, 2018, 71–87) and techniques such as machine learning (ML), deep learning (DL), natural language processing (NLP), robotics, speech, computer vision, supervised learning, and unsupervised learning (for more information, see Table 10.1).

AI will be utilised for processing the massive amounts of intelligence, surveillance, and reconnaissance (ISR) data involved in modern operations. Furthermore, the military applications of AI will be manifested in the development of autonomous vehicles. AI-based guidance systems will support space and undersea platforms, while the so-called drone swarms, will also be driven by AI (National Academies, 2018). AI it is anticipated that will enrich battlefield simulations and war games to explore dynamic conditions (weapons, allies, etc.) and their impact on decision-making, analyse games, and gather intelligence via satellite, drones and cyber domain (Reddie et al., 2018, 1362–1364). One, however, should not underestimate the fact that western rivals such Russia and China will also have the potential to change the strategic calculus with the use of AI in military domains as well (Crosston, 2020).

Cyber space is the global domain within the information environment consisting of the network of information technology infrastructures and data, including the

Table 10.1 Key Applications in AI

Deep Learning	The most complex forms of machine learning (ML) involve *deep learning (DL)*, or neural network models that comprise many levels of features or "variables" that can predict outcomes. This ML-based approach utilises a logic structure similar to the brain called neural "networks" to recognise and discriminate patterns such as speech, image and video (Properzi et al., 2019) While most ML programs can work with small data sets that are organised and labelled, DL programs are most effective when applied to large volumes of raw and unstructured data (Mollica et al., 2016). DL techniques are also increasingly being used for speech recognition and, as such, this form of analysis is becoming embedded in NLP systems.
Natural Language Processing	This refers to the application of computational techniques aimed at analysing and synthesising natural language and speech; and includes applications such as speech recognition, text analysis, translation and other goals related to language. The objective of NLP is not only to establish the structure between words in a text (syntax), but to also to understand the meaning (semantics) and the context meaning (pragmatics) (Rueda et al., 2019)
Machine Learning	Machine learning is a statistical technique for fitting models to data and to "learn" by training models with data. It is one of the most common forms of AI and includes various technologies such as DL, supervised learning, unsupervised learning, and reinforcement learning. ML uses computer algorithms that learn from structured and unstructured data to identify hidden patterns, make classifications and predict future outcomes.

Internet, the World Wide Web Information System, telecommunications networks, computer systems, and embedded processors and controllers. In today's world, cyber space transcends geographical and geopolitical boundaries and is critical for commerce, governance, and national security. Along with the physical domains of air, ground, maritime, and space, cyber space is one of the five interdependent spheres of human operations. Modern operations in the air, on land, and at sea depend on information technology infrastructures and computers, as well as the information that flows and is processed in cyberspace. Furthermore, all space domain operations are inextricably connected with cyberspace, and certain cyberspace operations cannot be carried out without the assistance of the space domain (e.g. satellite applications and communications).

Between 1960s and 1980s, both NATO and Soviet Union had launched numerous military missions to address requirements for space reconnaissance, telecommunications, navigation, weather monitoring, command, and control while simultaneously developed their offensive intercontinental ballistic missile (ICBM) capabilities and their space infrastructure, symbolised by the Program Apollo and the numerous Space Stations from both sides.

Despite all the Cold War developments, space never dominated the theatre of operations until Gulf War in 1991, the first ever "Space War." Global positioning system (GPS) would change the warfare and space-based navigation, imaging, and communications would become indispensable assets for all warfighters. In the three decades after the Gulf War, we have seen the establishment of new corps like the US Space Force and the UK Space Command, the establishment of mass production lines for satellites (e.g. Airbus One Web Satellites, over 7,000 Starlink satellites in low Earth orbit) but we have also seen a dramatic increase of overt testing of anti-satellite weapons.

As the contours of new technologies become increasingly clear over time, the question on how military leaders and training will be able to adapt and embrace these technologies remains to be seen. People are an organisation's most valuable asset, and the most successful companies are those in which the leadership drives and utilises human capital most effectively and efficiently. The expertise, skills, experience, and education that each employee possesses and applies inside the organisation determine the value of this human capital. This realisation points to a specific type of leadership characteristics which can be identified as the transformational type and will be assessed in the following section.

Leadership and Soldier Development in the Age of High-Tech Warfare

The prevailing view among academics sees leadership as a social process in which one individual shape the attitudes, values, and, most importantly, actions of others (Yukl, 2013). Consequently, leadership development has become increasingly important for businesses and organisations of all sizes (Hotho and Dowling, 2010) including militaries, even though the type of training program may vary by company or sector.

Military leadership is a distinct theoretical and operational discipline from civilian leadership. The framework that defines military leadership is, primarily a military organisation's core duty, which is to provide security. Members of the armed forces are therefore, given the authority to use force to fulfil this responsibility (Kart et al., 2016, 159–187). Military leadership is severely influenced by the combat and security missions it performs. Military organisations are also known for their "totality," in which regulate nearly every part of their members' lives, while another key characteristic of the military organisation is that military forces aim to accomplish their objectives through a hierarchical structure. Due to the organisation's scale, commanders' decisions and actions significantly impact a large number of subordinates.

In addition to the organisational context, understanding military leadership requires a study of the operational environment in which it functions, as well as the major social, cultural, and technological changes that shape it (Morath et al., 2011, 453–461). Today's modern militaries are dealing with a "spectrum of operations and confrontations" which makes it imperative that they must have the ability to function efficiently in environments that quickly shift from peace to outright war. For example, military forces in Iraq and Afghanistan had to simultaneously operate and fight in urban environments while also providing humanitarian assistance (Konaev, 2019). In this ever more complex and unpredictable environment, military leaders must be trained to carry out a variety of operations. They must be able to instantly assess the environment, take actions, devise strategies, and adapt to unexpected outcomes. It is not then a surprise, that versatility, agility, adaptability, flexibility, ingenuity, and the drive and ability to participate in continuous learning are all necessary skills for success in today's (and tomorrow's) operational environments, according to US Army leadership doctrine (Morath et al., 2011)

Since military leaders operate within a hierarchical structure, they are not only responsible for the success of their own organisation but also bear responsibility for the broader society in which their organisation functions. Society, like members of the armed forces, expects military leaders to be both efficient and attentive to the well-being of their troops. Military leaders are often seen as integral members of the broader political system. They represent the military institution, giving it both a face and a voice, and shape its dominant leadership culture through their actions and decisions (*Army Doctrine Publication, 2019*).

When a military leader's level of command rises, so does the complexity of his or her decision-making. Studies show that when leaders involve skilled subordinates in the decision-making process, adherence to decisions and effective implementation improve significantly (Ejimabo, 2015).

While desirable leadership traits have evolved over time, the basic formula for leadership success has remained relatively unchanged for the past 2,000 years (Table 10.2). The method for instilling, encouraging, and maintaining the required leader attitudes, on the other hand, has yet to be determined (Walter, 2010).

Table 10.2 Key Characteristics of Military Leadership

Discipline	Military leadership is a distinct theoretical and operational discipline from civilian leadership, which includes both normative and context-specific elements.
Mission Context	Military leadership is severely influenced by the combat and security missions it performs.
Regulation	Military organisations are also known for their "totality," in which regulate nearly every part of their members' lives.
Hierarchy	Military forces aim to accomplish their objectives through a hierarchical structure.
External environment	Operational theatre in which military leadership currently acts as well as the major social, cultural, and technological changes that characterise it.
Decisiveness	Leaders must have the ability to function efficiently in environments that quickly shift from peace to outright war.
Versatility	Leaders must be able to instantly assess the environment, take actions, devise strategies, and adapt to unexpected outcomes.
Value of Education	Leaders must be able to participate in continuous learning initiatives.
Social Responsibility	Leaders are largely responsible not only for the success of their own organisation, but they bear responsibility for the entire society in which their organisation is part of it.

Evolving Command: How 21st-Century Challenges Shape Military Leadership

Military leadership is often seen as encompassing two key elements: task-oriented leadership (transactional leadership) and change-oriented leadership (transformational leadership) (see Meerits and Kivipõld, 2020). James MacGregor Burns' 1978 book Leadership is widely regarded as a pivotal work in introducing a new approach to leadership (Burns, 1978). According to Burns, the basic role of leadership is to bring the leader's and subordinates' goals and objectives together in order to achieve a higher vision. This way of thinking implies the possibility that people do not have to agree on anything, but that their shared vision and goals must bring them together.

Burns's core premise was to distinguish between two types of leadership (the transactional and transformational styles (McCleskey, 2014, 117–130). The most common form of leadership is a transactional leadership. It is centred on mutual activity, which occurs when a leader approaches a subordinate to share something, but it is important in transactional leadership that the leader seeks to achieve organisational goals and objectives by controlling his subordinates. However, transformational leadership is more challenging and useful than transactional leadership (Diaz-Saenz, 2011). A leader considers and focusses on the wishes and demands of potential subordinates in this case. Furthermore, a transformational leader seeks to understand his subordinates' motivations, to meet their needs, and secure their commitment.

Studies have also identified the necessary skills for 21st-century leaders that can be applied to various sectors including the military. They have usually included the ability to cope with cognitive uncertainty, intellectual agility, a significant degree of self-awareness, and a better understanding of the relationships between organisational sub-systems. These characteristics can be added to the classic qualities of a leader: honesty, energy, bravery, and adherence to institutional ideals (Walter, 2010).

In exploring how to develop modern military leaders, the concept of "best practice" emphasises the importance of strong, supportive relationships between leaders and soldiers. In industries, the idea of "best practice" is regularly revised and defined but what works universally remains uncertain. Given the disparities between civilian and military organisations, but also mindful of the shared characteristics that all big, complex organisations share, one can identify the following important elements that are important in the development of military leadership in 21st century.

One is that attracting, motivating, and developing high-quality leaders needs a welcoming, fair work environment. Another is that the principles, knowledge, and attitudes of the organisation's senior leadership have a significant impact on organisational climate and cohesion (Jans and Schmidtchen, 2002, 1–204).

Military leaders, more than political and business leaders, are confronted today with an unprecedented complex warfare environment. In addition to the leadership challenges presented by the complexity and ambiguity of modern warfare (Hoffman, 2007), the conflicts in Iraq, Afghanistan, and Libya, have highlighted the critical need for cultural awareness (Nohria and Khurana, 2010). To make matters worse, alliance and coalition partners' cultures can also differ to varying degrees and might have a significant impact on the outcome of multinational operations. It does not then come as a surprise that the US Army makes sure that soldiers are trained to perform in any *volatile, uncertain, complex, and ambiguous* (VUCA) situation (Lawrence and Steck, 1991).

Despite the range of challenges and responsibilities that today's military leaders and soldiers face, some of which are long-standing while others are new, several core roles remain essential for a 21st-century military leader.

Warrior-Leader: Officers and non-commissioned officers must continue to lead men and women into danger. They must plan, prepare, and lead their units in a variety of missions.

Caretaker/Guardian: Decades of conflict including the twenty-year war in Afghanistan and significant coalition losses in Iraq have put a tremendous strain on forces and their families. Military leaders must ensure the physical and mental well-being of service members and their families.

Caretaker/Guardian of Institution: Leaders are committed to developing conditions that promote the learning, growth, and retention of service members in time of war and peace.

Technical Experts: Leaders must acquire and retain the technological and tactical skills needed to lead across an ever-broadening spectrum of missions and operational environments. As they lead this highly dispersed army, leaders must also be skilled in the use of communication technology (Dubois et al., 2017).

The advancement of technology has become one of the most powerful drivers of transformation in the military operating environment. The NATO countries (as well as potential western adversaries and rogue states) have significantly improved their capabilities thanks to technological developments in military equipment and systems (McDonald, 2021).

Ultimately, military leadership is similar to leadership in any other profession or occupation in terms of priorities and the need to complete tasks. There are differences in terms of context—where the leadership takes place (Chan et al., 2011). Military leaders are planning for or working in serious and extreme conditions marked by difficulty and ambiguity, which may include using or being attacked by lethal force (Wong et al., 2003).

From the above analysis and the point of view of military leadership, the type of transformational leadership seems to be the most relevant to lean on in the 21st century (Nissinen, 2001). In both military and commercial environments, "transformational" leaders have been found to be more successful than leaders who rely heavily on transactional leadership style (Bass and Avolio, 1993). The Ukraine war provides a powerful example of how transformational leadership can succeed in the face of overwhelming odds. Commanders like General Valerii Zaluzhnyi have exemplified this leadership style by articulating a clear vision of national defence, inspiring their troops, and fostering a culture of innovation and resilience. Furthermore, academics such as Burns argue that transactional leadership activities lead to short-term trading relationships between followers and the leader. These relationships are characterised by superficial, intermittent exchanges that often result in dissatisfaction among the participants (Burns, 1978). Several academics criticise transactional leadership theory for adopting a generic, one-size-fits-all approach to leadership that overlooks situational and contextual factors critical to 21st-century military operations and organisations (Beyer, 1999). Although empirical studies often support transactional leadership, they typically examine it alongside transformational behaviours (Gundersen et al., 2012, 46–57).

Forging Tomorrow's Military Leaders

New technology has transformed 21st-century warfare, and the rapid pace of these advancements will challenge leaders, soldiers, and decision-makers (Latiff, 2017; Cohen et al., 2020). The Ukraine conflict underscores the necessity of preparing leaders for asymmetric and hybrid threats. The rise of hybrid threats (Fleming, 2011) as well as the proliferation of regional conflicts (the Russia–Ukraine War, the Israeli–Hamas/Hezbollah conflict, Libyan conflict of 2011, and the Syrian civil war) and asymmetric warfare engagements (including operations in Afghanistan,

Mali, or against global extremist movements such as Daesh), requires a shift not only in strategic and tactical approaches but also in cultural awareness.

Cultural awareness enhances the ability to communicate effectively with individuals from various backgrounds, which is crucial for building trust and fostering cooperation. Additionally, Ukraine's ability to integrate Western military aid into its operations demonstrates the importance of coalition leadership. At all levels, leaders must recognise, appreciate, and accommodate cultural differences in order to build long-term relationships and cooperation that support the mission's success.

Since technological systems provide an unprecedented amount of information to units and commanders at all levels, leaders should be trained to sift through a constant stream of data to find and synthesise the relevant facts into a coherent picture of the situation. A leadership training program based on transformational leadership requires that each military leader strive for personal development and growth. Meanwhile, as military operations become ever more complex, the need to instil a *culture of innovation, adaptability, and agility of decision-making* in the training of future leaders will be unavoidable if not a panacea.

Modern military organisations should foster a culture of continuous education and learning by:

- Expanding negotiation, coaching, and mentoring opportunities at all levels, particularly at junior ranks.
- Emphasising a commitment to personal and professional growth from the outset of one's career, sustaining it throughout.

For instance, the U.S. military offers targeted language training to enhance regional understanding and routinely sponsors both enlisted personnel and officers for degree and postgraduate studies in relevant fields such as cryptology, international relations, and strategic studies (Ellinger and Posard, 2023).

The "freethinkers" do not get promoted because they are not conventional enough. This is not helped by the requirement to do specific jobs/roles prior to promotion for certain amounts of time. A risk adverse culture has been grown in militaries. There needs to be more incentives for officers to think "outside the box." We train and educate military people to be risk adverse (i.e. engineering officers); they therefore lose their agile thinking.

Adopting the transformational leadership model requires militaries to meet the challenges of the twenty-first century by educating, engaging and utilising their general officers' attitude, knowledge and critical thinking skills. A good example of this approach can be demonstrated by the United States Army, which in 2010 formally adopted the idea of "design" to address the "fog of war" scenarios into its operational planning doctrine (Department of the US Army, 2010). The idea stemmed from a growing awareness that the traditional planning process had not delivered the level of understanding needed in the contemporary operational context, and that the why-to-what component of operational planning required more systematic review. The idea of design was instigated prior to the planning of major military operations, a design team was to be constituted in order to properly understand the dynamics of:

i the operational environment;
ii the problem at hand;
iii the potential operational approaches available.
iv Once those dynamics had been understood, the commander might then synthesise a concept of operations that could be handed down to the planning team, where it would inform their planning of subsequent force generation, manoeuvre, fires, logistics, etc. Utilising the subordinates' critical thinking skills, knowledge, desire to participate in the mission, will only benefit the organisation and its members.

As new technologies become central to military operations, those tasked with integrating these advancements must not lose sight of the human component in training and education. Training should not merely be an afterthought, but it should be integrated into development.

Emerging domains like space, cyber, AI will require both academic but also industrial training to understand the nature of these systems. Universities and the defence industry have significant experience with these technologies, and they need to work very closely with the future military leaders to make them aware and educate them on these innovations and their capabilities.

In addition to standardised training, these new capabilities must be continuously integrated into war games, exercises, and demonstrations to improve not only the operators' performance but also the ability of the military leaders to incorporate these capabilities into campaigns. Demonstrations and simulations are critical to evaluate the functionality, requirements but most importantly to get a sense of what is needed in terms of further training requirements. Ideally, these simulations need to incorporate operational concept demonstrators where the military leaders and the armed forces will be able to evaluate these capabilities in the real theatre of operations.

In the complex/hybrid environment of the 21st century, the military will require more agility, forward deployment capabilities, better situational awareness, fully integrated information operations, and sustained operational persistence. Achieving those things will involve more devolution of command and control, better-integrated information systems at the alliance and joint levels and well-educated officers.

Conclusion

This chapter has demonstrated that leadership needs to evolve in response to significant transformations in the military environment. The rise transformative technologies will challenge the ability of military leaders to adapt to an increasingly dynamic and complex battlefield. By understanding how these technologies influence future operations, western military leaders can better prepare, organise, and strategize for upcoming missions.

As leaders continue to learn how to lead and serve the members of their organisations, new technologies are pushing them to retool. Additionally, the lessons NATO armies have gained from recent operations in Ukraine, Iraq, Afghanistan, Syria, and Libya should also be used to revise national strategies for military leadership.

As highlighted in the chapter, a leadership-training program based on transformational leadership will require that each military leader strive for personal development and growth. Simultaneously, the need to instil a culture of innovation, adaptability and agility of decision-making in the training of future leaders will be unavoidable if not a panacea.

Military leadership in 21st century should embrace officers desire to further themselves. Conflicts and wars take place between people. People are influenced in two ways by leadership training: directly and indirectly. We have the potential to shape our leadership culture because a military institution has the ability to educate its leaders. Only learning organisations will thrive in the world of change and innovation.

References

Army Doctrine Publication (ADP) 6-22. 2019. "Army Leadership and the Profession." July 31. https://armypubs.army.mil/ProductMaps/PubForm/Details.aspx?PUB_ID=1007609 (accessed May 20, 2021).

Avolio, B. J. 2010. "Pursuing Authentic Leadership Development." In *Handbook of Leadership Theory and Practice: A Harvard Business School Centennial Colloquium*, edited by N. Nohria and R. Khurana, 739–768. Boston, MA: Harvard Business School Publishing.

Bass, B., and Avolio, B. 1993. *Improving Organizational Effectiveness Through Transformational Leadership*. Thousand Oaks, CA: Sage.

Beyer, J. M. 1999. "Taming and Promoting Charisma to Change Organizations." *The Leadership Quarterly* 10 (2): 307–330. https://doi.org/10.1016/S1048-9843(99)00019-3.

Burns, J. M. 1978. *Leadership*. New York: Harper & Row.

Chan, K.-Y., Soh, S., and Ramaya, R. 2011. *Military Leadership in the 21st Century: Science and Practice*. Singapore: Cengage Learning Asia.

Cohen, R., et al. 2020. *The Future of Warfare in 2030*. Santa Monica, CA: Rand Corporation. www.rand.org/pubs/research_reports/RR2849z1.html (accessed May 20, 2021).

Crosston, M. 2020. "Cyber Colonization: The Dangerous Fusion of Artificial Intelligence and Authoritarian Regimes." *Cyber, Intelligence and Security* 4 (1): 149–171.

Davis, Z., and Nacht, M., eds. 2018. *Strategic Latency: Red, White and Blue, Managing the National and International Security Consequences of Disruptive Technologies*. Berkeley, CA: Lawrence Livermore National Laboratory, 71–87.

Diaz-Saenz, H. R. 2011. "Transformational Leadership." In *The SAGE Handbook of Leadership*, edited by A. Bryman, D. Collinson, K. Grint, B. Jackson, and M. Uhl-Bien, 299–310. Thousand Oaks, CA: Sage.

DuBois, R. F., Gerstein, D. M., and Keagle, J. M. 2017. *Science, Technology, and U.S. National Security Strategy: Preparing Military Leadership for the Future*. CSIS Reports.

Ejimabo, N. O. 2015. "An Approach to Understanding Leadership Decision Making in Organizations." *European Scientific Journal* 11 (11): 24.

Ellinger, E., and Posard, M. N. 2023. *Imagining the Future of Professional Military Education in the United States*. RAND Corporation.

Fleming, B. P. 2011. "Hybrid Threat Concept: Contemporary War, Military Planning and the Advent of Unrestricted Operational Art." Monograph. US Army Command and General Staff College.

Gundersen, G., Hellesoy, B. T., and Raeder, S. 2012. "Leading International Project Teams: The Effectiveness of Transformational Leadership in Dynamic Work Environments." *Journal of Leadership & Organizational Studies* 19 (1): 46–57. https://doi.org/10.1177/1548051811429573.

Hoffman, F. G. 2007. *Conflict in the 21st Century: The Rise of Hybrid Wars*. Arlington, VA: Potomac Institute for Policy Studies. https://potomacinstitute.org/reports/19-reports/1163-conflict-in-the-21st-century-the-rise-of-hybrid-wars (accessed May 20, 2021).

Hotho, S., and Dowling, M. 2010. "Revisiting Leadership Development: The Participant Perspective." *Leadership & Organization Development Journal* 31 (7): 609–629.

Jans, N., and Schmidtchen, D. 2002. *The Real C-Cubed: Culture, Careers and Climate and How They Affect Military Capability*. Canberra Papers on Strategy and Defence, no. 143. Australian National University.

Jensen, B., et al. 2020. "Algorithms at War: The Promise, Peril, and Limits of Artificial Intelligence." *International Studies Review* 22: 526–550.

Kark, R., Tair Karazi-Presler, and Sarit Tubi. 2016. "Paradox and Challenges in Military Leadership." *Leadership Lessons from Compelling Contexts: Monographs in Leadership and Management*. Bingley, UK: Emerald Group Publishing, 159–187.

Konaev, M. 2019. "The Future of Urban Warfare in the Age of Megacities." *Focus Strategique*, no. 88. Ifri, March. www.ifri.org/en/publications/etudes-de-lifri/focus-strategique/future-urban-warfare-age-megacities (accessed May 20, 2021).

Latiff, R. 2017. *Future War: Preparing for the New Global Battlefield*. New York: Alfred A. Knopf..

Lawrence, J. A., and Steck, E. N. 1991. *Overview of Management Theory*. Carlisle Barracks, PA: U.S. Army War College.

Marr, B. 2018. "The Key Definitions of Artificial Intelligence (AI) that Explain its Importance." *Forbes*, February 14. www.forbes.com/sites/bernardmarr/2018/02/14/the-key-definitions-of-artificial-intelligence-ai-that-explain-its-importance/#5b0977914f5d (accessed April 29, 2021).

McCleskey, J. A. 2014. "Situational, Transformational, and Transactional Leadership and Leadership Development." *Journal of Business Studies Quarterly* 5 (4): 117–130.

McDonald, J. 2021. "Remote Warfare and the Legitimacy of Military Capabilities." *Defence Studies*. https://doi.org/10.1080/14702436.2021.1902315.

Meerits, A., and Kivipõld, K. 2020. "Leadership Competencies of First-Level Military Leaders." *Leadership & Organization Development Journal*. www.emerald.com/insight/0143-7739.htm (accessed April 29, 2021).

Mollica, L., Decherchi, S., Zia, S. R., Gaspari, R., Cavalli, A., and Rocchia, W. 2016. "Kinetics of Protein-Ligand Unbinding via Smooth Potential Molecular Dynamics Simulations." *Scientific Reports* 6. www.nature.com/articles/srep11539 (accessed June 7, 2021).

Morath, R. A., Leonard, A. L., and Zaccaro, S. J. 2011. "Military Leadership: An Overview and Introduction to Special Issue." *Military Psychology* 23 (5): 453–461.

National Academies of Sciences, Engineering, and Medicine. 2018. "Counter-Unmanned Aircraft System (CUAS) Capability for Battalion-and-Below Operations." Abbreviated Version of a Restricted Report. www.nap.edu/catalog/24747/counter-unmanned-aircraft-system-cuas-capability-for-battalion-and-below-operations (accessed May 29, 2021).

Nissinen, V. 2001. *Military Leadership: Critical Constructivist Approach to Conceptualizing, Modelling and Measuring Military Leadership in the Finnish Defence Forces.* Helsinki: National Defence College.

Nohria, N., and Rakesh, K., eds. 2010. *Handbook of Leadership Theory and Practice.* Harvard Business Press.

Properzi, F., Taylor, K., Steedman, M., Ronte, H., and Haughey, J. 2019. "Intelligent Drug Discovery." *AI: Deloitte Centre for Health Solutions.*

Reddie, A. W., et al. 2018. "Next Generation War Games." *Science* 362 (6421): 1362–1364.

Rueda, J. D., Cristancho, R. A., and Slejko, J. F. 2019. "Is Artificial Intelligence the Next Big Thing in Health Economics and Outcomes Research?" *Value and Outcomes Spotlight*, March/April. www.ispor.org/docs/default-source/publications/value-outcomes-spotlight/march-april-2019/vos-heor-articles---rueda.pdf?sfvrsn=18cb16f5_0 (accessed April 30, 2021).

U.S. Army. 2010. *FM 5-0 The Operations Process.* Department of the U.S. Army.

Ulmer, W. F., Jr. 2010. "Military Leadership into the 21st Century: Another Bridge Too Far." *Parameters* 40 (4): 138.

Wong, L., Bliese, P., and McGurk, D. 2003. "Military Leadership: A Context Specific Review." *The Leadership Quarterly* 14: 657–692. https://doi.org/10.1016/j.leaqua.2003.08.001.

Yukl, G. 2013. *Leadership in Organizations*, 8th ed. Upper Saddle River, NJ: Pearson Education Inc.

11 Ukrainian Tactical Innovations during Russia's Full-Scale War Against Ukraine, from February 2022 to September 2024

Lieutenant Colonel (LtCol) Daniel Love

Introduction

The focus of this chapter is on selected innovations undertaken by Ukrainians in Russia's full-scale war against Ukraine since 24 February 2022 and until September 2024. The developments and innovations in the conduct of this war have been rapid and myriad. This chapter is an attempt to capture selected innovations, specifically regarding military operations at the tactical level in the conduct of this war. Ukraine is fighting an existential war against an unprovoked and illegal invasion by the Russian Federation (RF). The key takeaways for Western military readers are how to better support its allies that may find themselves, like Ukraine, fighting an existential war against a hostile invader, as well as how to prepare their own militaries for both the verities and the game-changing innovations of 21st-century warfare that are being evidenced on the battlefield in Ukraine over the last two and a half years. This chapter is not all-encompassing or exhaustive, and its scope is to provide an overview of interesting developments in order to serve as a jumping-off point for further study and research on a given topic of interest.

This chapter is likewise based on reporting, research reports, and books from 2022 to present. Most of these sources are based on accounts from Ukrainian military service members, or Ukrainian sources otherwise, with first-hand knowledge of events that they have been shared with journalists, researchers, and historians. These writings are not only a first draft of history of an ever-evolving and highly dynamic situation in Ukraine's war of survival, but also ones that offer insights from which key takeaways and lessons for this war and for war in 21st century can be gleaned. This chapter is an attempt to provide relevant references from the aforementioned reporting, research reports, and books and provide analysis in the form of key takeaways, implications, and conclusions regarding this conflict as well as war in the 21st century.

Tactical Level Innovations and Developments

Ukraine's war of survival is one in which Ukraine has fought with resource and manpower shortfalls, comparative disadvantages with the RF's resources and

DOI: 10.4324/9781003520160-11

manpower, as well as rules of engagement constraints imposed by the U.S. government. Despite these shortfalls, comparative disadvantages, and constraints, Ukraine has succeeded in over two and a half years of large-scale combat operations (LSCO), holding a much larger, much better-resourced invader at bay and destroying multitudes of RF resources, specifically military equipment, personnel, and war materiel in the process (General Headquarters Armed Forces of Ukraine 2024). Due to Ukraine's success, during the course of this war, Russia has reassessed and adjusted its strategic aims at least four times to date in order to align its end state with what it might be able to achieve tactically and operationally in its war of aggression against Ukraine (Freedman 2023, 44–45). Two areas at the tactical level of war that evidence innovation and that have contributed to Ukraine's success thus far are unmanned aerial systems (UAS) and artillery operations. The following section will explore the innovations and adaptations in these areas, primarily from the Ukrainian side, in order to demonstrate their impact on this conflict's progression as well as to offer key takeaways, implications, and conclusions for this conflict and future wars in the final section.

Unmanned Aerial Systems (UAS)

The quantity of UAS employed by both Ukraine and the RF on the battlefield in this war is without precedent, as are the types of UAS being used at scale and the methods in which they are being employed. Prior to Russia's full-scale war against Ukraine, UAS typically referred to expensive fixed-wing platforms that conducted Intelligence, Reconnaissance, and Surveillance (ISR) missions or served as a means to conduct a lethal weapons strike, usually operated by the U.S. or other large militaries. In this conflict, while fixed-wing platforms for ISR or lethal strikes are still employed, UAS typically means ubiquitous drones, most often of the quadcopter variant, and often ones that are lethal, cheap, and disposable. The tactics developed, refined, and currently used by Ukraine are changing the way this war, and future wars, are fought.

The most prominent change regarding UAS platforms has been the ability to deliver lethal munitions strikes to personnel, equipment, or war materiel using comparatively cheap drone systems, resulting in disproportionate and highly cost-effective kills. These drones not only cost exponentially less than the materiel they destroy, but also can be replaced much quicker than their targets. The U.S. Army War College's *A Call to Action: Lessons from Ukraine for the Future Force* provides an illustration of the evolution of employment of drones in Ukraine's war. The authors write:

> A squad of Russian soldiers crouches in the eastern Ukrainian woods late in the evening. Suddenly, a buzz overhead sends four soldiers running, while three remain stationary. A Ukrainian soldier at an unknown distance watches the scene through a screen, beamed back to him in infrared from the drone that has spotted the Russian troops. Zooming in on the three stationary soldiers,

> the Ukrainian drone stabilizes and drops a small munition into their midst. The Ukrainian military would later report all three Russian soldiers had been killed.
>
> Scenes like this have played out countless times on both sides of the current conflict in Ukraine. The United States and its allies are well practiced at delivering strikes from large UAVs circling high above the battlefield, but this scenario is different. Smaller, lighter, faster, cheaper, disposable, reconfigurable on a kitchen table, and readily accessible on electronic store shelves, this new version of the threat poses a challenge the United States has not seen previously. Export restrictions around unmanned aerial systems (UASs) have been relaxed, and the market has expanded, meaning UASs are not just the tools of advanced militaries.
>
> (Holbrook 2024, 197)

This vignette, referencing an event in the first three months of the war, was only the beginning. Countless other examples of lethal quad-copter drones have appeared on Telegram channels and X (formerly Twitter) feeds that cover Russia's war of aggression against Ukraine since, evidencing the innovative drone tactics being used in this war. Cheap drones, modified to be able to carry lethal munitions have become the apex predator on the battlefields in Ukraine. In September 2024, the Armed Forces of Ukraine (AFU) began experimenting with first-person view (FPV) quad-copter drones modified and able to fire either an attached rocket-propelled grenade (RPG) or anti-tank grenade launcher (AT), which will ostensibly lead to further adaptations in the tactics, employment, and formations of personnel and equipment by the RF on the battlefield once these RPG drones and AT drones are fully operationally capable (NEXTA 2024; ArmyInform 2024).

As of September 2024, FPV drones have been modified by the AFU to be able to drop hand-grenades or even larger payloads up to 9.5 kilograms (i.e. frag drones/bomber drones or "Queen Hornet"), fly rigged with explosives into a target (i.e. kamikaze drones/suicide drones), drop thermite on concealed RF positions (i.e. dragon drones), fly with a machine gun able to fire on enemy positions (i.e. machine-gun drones), fire rocket-propelled grenades (i.e. RPG drones), and fire anti-tank rockets (i.e. AT drones). While the latter two drone variations are still being tested by the AFU (NEXTA 2024; Wild Hornets 2024d), as of September 2024 there are verifiable instances of the former four drone variations being employed on the battlefield by the AFU (Wild Hornets 2024a, b, c). The Ukrainian start-up Wild Hornets, which came in to being in the spring of 2023, produces approximately 100 drones a day, and is at the forefront of innovating modifications that have enhanced quad-copter drones' capabilities to deliver effects on the battlefield (Urbancik and Glushko 2024).

> The organisation manufactures different FPV drones. Their "Standard Wild Hornets" model can achieve up to 160 kilometres per hour and carry payloads of 1.5 to 3 kilograms, primarily for so-called "kamikaze missions." These drones, also known as loitering munitions or "suicide drones," are unmanned aerial systems that can loiter for extended periods before engaging targets with built-in

> warheads, effectively combining the characteristics of precision missiles and UAVs. These drones are generally only used once.
>
> They also create so-called "bomber drones," which can be used several times, as well as the larger "Queen Hornet" model, which can carry up to 9.5 kilogram bombs and boasts a range of 30 kilometres. These drones are also used for operations like supplying food deliveries to front line areas. According to Forbes, both Ukraine and Russia are increasingly utilising drones for logistical purposes, with Ukrainian forces repurposing small FPV drones not only for combat but also to deliver essential supplies for their military personnel on the frontline.
>
> (Urbancik and Glushko 2024)

These innovative drone capabilities have forced armoured personnel carriers and tanks to change their employment tactics, which likewise have been outfitted with additional armour, overhead netting, and other protective measures, such as electronic warfare measures, in attempts to increase their survivability against frag, "Queen Hornet" bombers, or kamikaze and suicide drones (Kirichenko 2024). They have also had an impact on the employment of infantry on the battlefield, as now an FPV drone operator is able to provide suppression of enemy infantry both in the open, in trenches, or in cover and concealment (i.e. basically anywhere the FPV drone operator can locate infantry personnel on the battlefield). Furthermore, these drones are being modified with the capability to deliver ever larger payloads, which in some cases are tailored lethal munitions payloads that are delivered with FPV precision on enemy targets and in other cases are logistics payloads that are delivered to friendly positions to provide resupply.

So, what does all this mean for the course of this conflict and the future of warfare? According to T.X. Hammes' report entitled "Game-changers: Implications of the Russo-Ukraine War for the Future of Ground Warfare," in this war, drones "most important function has been to provide critical intelligence and artillery spotting for tactical units in contact with enemy forces" (Hammes 2023, 9). The evolution in drone warfare up to this April 2023 report was the pervasive and ubiquitous presence of small, hand-carried drones of mostly commercial origins that were used at all echelons of military formations in this conflict, particularly at the platoon, company, battalion, regiment, and brigade levels. This function of aerial observation for intelligence collection, ISR, and fire observation and adjustment by drones early in Russia's war of aggression against Ukraine is similar to the role played by manned aviation in its first forays on the battlefield during World War One (WWI). If this were the extent of the evolution of drone warfare during this conflict, such a development would have been revolutionary in itself.

However, the role of drones has quickly evolved in the course of this war. According to a GLOBSEC report entitled "How to Beat Russia: What Armed Forces in NATO Should Learn from Ukraine's Homeland Defense," Nico Lange asserts that "at the beginning of the war, for example, improvised commercial drones capable of releasing simple grenades by remote control were still derided as dubious gimmicks, they have since become an integral part of warfare against trenches and certain types of enemy vehicles" (Lange 2023, 21). The report goes on to note that "The leadership of the Armed Forces of Ukraine was quick to recognize the military significance of ubiquitous drone use and quickly adapted" (Lange 2023, 21). The AFU has subsequently established a centralised drone school to train its forces on UAS operations, providing a broad overview of the types of UAS platforms and their purpose in support of Ukraine's war efforts (Lange 2023, 21). Furthermore, the AFU has not only stood up as a centralised training school, given the importance of drones and UAS in this war, they are also experimenting and innovating with the integration of these trained drone and UAS operators at all echelons of their tactical military formations, including creating separate stand-alone UAS regiments to support combat operations. In just over two and a half years, drone warfare has evolved rapidly. First these platforms were employed for pervasive and ubiquitous aerial observation, intelligence collection, and artillery fires spotting and adjustment at the tactical level, but have subsequently evolved to allow for the delivery of lethal and relatively low-cost strikes against enemy troops and equipment. These emerging and evolving weapons of war allow for the cost-efficient disabling or destroying of multi-million dollar military equipment with a disposable and replaceable drone system that costs several hundred to a few thousand dollars at most (Urbancik and Glushko 2024; Hammes 2023, 9). Likewise drones, particularly fixed-wing platforms with extended ranges that are either ISR platforms used to facilitate lethal fires or platforms that have lethal strike capabilities themselves, are consuming the enemies' military materiel by requiring the expenditure of costly counter-measures (i.e. anti-air defense fires to take down these comparatively cheap but lethal drones).

In summary, Russia's war in Ukraine has evidenced the rapid evolution of drones in modern warfare, and what will likely come to pass before the end of this conflict and in future wars is quite problematic for Russia, Ukraine, and any future belligerents in LSCOs. In his book *Modern Warfare: Lessons from Ukraine*, Lawrence Freedman sums up the role of drones in Russia's war of aggression against Ukraine:

> The use of drones has been one of the more innovative aspects of the war. Both sides have used them extensively, although Ukraine was better prepared … Commercial UAVs became progressively more important to both sides. They were relatively cheap and relatively easy to acquire and adapt, and therefore expendable. The Ukrainians learned to arm them with grenades and use them over short distances against unsuspecting Russian troops … . The Ukrainians also produced strike drones, able to deliver bombs and missiles over long distances, including into Russian territory. The use of drones indicates the importance of innovation and adaptation during a war. Both sides embraced simple but workable solutions rather than relying on only the most capable, high-performance systems.
>
> (Freedman 2023, 99–100)

Quite simply, Ukraine has made drones efficient, cost-effective, relatively cheap, readily available, and lethal at all tactical echelons. The AFU are using drones in tactical engagements to conduct vertical envelopment of RF troops and equipment, more often than not killing or severely wounding personnel and degrading or completely destroying war equipment and materiel with cheap drones. This has provided the AFU an economy of force option that has alleviated some of the disadvantages it faces against Russia, given the former's comparative shortfalls in personnel, equipment, and war materiel (NATO PfPC 2023, 72). More importantly, drones have proven effective on the battlefield against RF forces and have also given Ukraine strategic depth to strike deep into Russian territory using domestically produced drones such as the Palianytsia (MacKay 2024). The RF forces and Ukrainians have developed some effective counter-measures, such as using electronic warfare (EW) effects to disable drones. However, these EW countermeasures also often degrade the employing sides' own capabilities and are also not sufficiently available at scale to protect all parts of a military formation against the ubiquitous drone threat posed by the current evolution in drone capabilities and pervasiveness.

These developments are likely the beginning stages of drones' roles and capabilities in war in the 21st century, much like aviation in WWI was only the beginning of its roles and capabilities in 20th century warfare. Mark Bowden predicts what may be the future of drone warfare, and it is grim:

> What might that future actually look like? For years, military strategists have anticipated the arrival of the so-called drone swarm, a large cluster of small flying machines that will herald a new era of intelligent warfare. Thousands of robotic aircraft no bigger than a starling would be all but invisible when spread out, yet capable of instantly coalescing into a swirling dark cloud, like a murmuration. It would move the way such phenomena move in nature, guided by a kind of group intellect...When you consider that a drone swarm consisting of many thousands of off-the-shelf drones would cost less than, say, one F-35 fighter or a ballistic missile, you have a weapon that would give rogue states or terrorist groups the means to launch devastating attacks or assassinations anywhere in the world. Since the Korean War, American forces have controlled the skies wherever they have gone into battle. No other nation had the means to compete with it; the cost, the technology, the experience, and the level of training required are beyond the reach of even the most affluent nation-states. Drone swarms could end that domination. An aircraft carrier? A commercial airliner? The White House? The president? Sitting ducks.
>
> (Bowden 2022)

Drones', and other UAS', roles and capabilities are rapidly evolving both within this war and likely in research and development (R&D) agencies in all leading global powers. The implications are that whichever global power develops and harnesses the capabilities of drones, artificial intelligence (AI) is able to control them and enhance their lethality, and associated supporting technologies will have a decisive advantage in conflicts in the 21st century. Airpower and air supremacy provided those who were able to develop the capabilities, formations, doctrine, and employment the advantage in 20th-century warfare. Drone warfare, with its ability to deliver persistent, cost-effective, continuous, and pervasive observation and lethal effects to the 21st-century battlefield could provide a similar advantage in 21st-century warfare for those who are able to further develop capabilities, formations, doctrine, and employment that are on display in their nascent stages in Russia's war of aggression against Ukraine.

Artillery Operations

There are many experienced artillerymen that are better qualified to analyse the implications of the changing, or in the case of mass of fires the enduring, nature of war regarding artillery given its use and evolution in Russia's war of aggression against Ukraine. However, in the course of researching developments in this war, a few salient innovations and developments in artillery operations seem pertinent to mention. Foremost is Ukraine's home-grown development of its "Kropyva" software to make allocation of fire missions both more expedient and efficient. Furthermore, the biggest takeaway from this conflict is that in neither this war nor future wars will mass of artillery fires become obsolete, as some have predicted since the advent of precision-guided munitions (PGMs) during the late 20th-century's revolution in military affairs (RMA). Lastly, and as a consequence of this enduring nature of mass of artillery fires, the importance of a nation's, and their allies', defense industrial base (DIB) production is a centre of gravity for this current conflict and future wars.

Regarding the use of the Ukrainian software Kropyva, foremost this is an example of how Ukraine is continuously innovating their tactics as well as resources available to them in this existential war in order to mitigate personnel and resources shortfalls and disadvantages. Ukraine started the war with no PGMs in their arsenal, and were at anywhere from a 10:1 to a 5:1 disadvantage in traditional artillery ammunition available when compared to the RF from the outset of hostilities and throughout 2022 and into 2023's well-publicised battle for Bahkmut (Freedman 2023, 57; Bradley III 2024, 98; Watling, Danylyuk, and Reynolds 2024, 12; Trofimov 2024, 335–337). Nico Lange notes that "it is indisputable that nearly all Ukrainian soldiers say that without the 'software weapon system,' Kropyva, it is unlikely they would be alive" (Lange 2023, 12). *The Wall Street Journal*'s chief foreign affairs correspondent, Yaroslav Trofimov, has written a highly informative account of the first year of Russia's war of aggression against Ukraine. In "Our Enemies Will Vanish: The Russian Invasion and Ukraine's War of Independence," he gives a first-hand account of the AFU's use of Kropyva in May of 2022 during combat operations in the village of Virnopillia, which is 12 miles southwest of Izyum in Kharkiv Oblast:

> We waited as Oleh peered through binoculars, making notes on his tablet loaded with Kropyva, the Ukrainian military's mapping software used to

> calculate artillery coordinates. He had spotted a Russian BMP [Boevaya Mashina Pekhoty, i.e. an infantry fighting vehicle] in the next tree line. The Russians were really close … "We're good to go," Oleh whispered loudly. "Stayed here long enough." We walked back through the forest for a minute or so, then the shelling began. We had been noticed … Back in the school basement, Oleh gathered with other commanders around his tablet, relaying the Russian BMP's location to an artillery unit. The strike would come within minutes.
>
> (Trofimov 2024, 200–204)

Trofimov's first-hand account reflects the use of Kropyva in action on the battlefield in Ukraine, noting how a volunteer battalion at the time on a secondary front of the AFU could still get mass fires support from traditional artillery formations within minutes, all thanks to this innovative software that networked all infantry units and forward observers and sensors otherwise with all available artillery weapons units and their systems. Kropyva's innovative networking has facilitated almost real-time artillery fire missions in the paradigm of "any sensor, best shooter" described by the U.S. Army War College's analysis on this same software.

The implications of this software are elaborated on by the authors of the U.S. Army War College's *A Call to Action: Lessons from Ukraine for the Future Force*. They state that Kropyva is the "most significant artillery modernization Ukraine has implemented" (Bradley III 2024, 101). This software, developed by Ukrainians, is a ballistic calculator application able to be downloaded and used on Android phones and tablets and "increased the functionality, accuracy, and response times of Ukraine's legacy artillery systems with software that resembles rideshare applications … earning it the moniker 'Uber for artillery' " (Bradley III 2024, 101). The authors note that:

> Soldiers with Android tablets can responsively deliver fires … Ukraine's frugally practical applications provide a real-time, handheld common operational picture that includes battlefield intelligence, which enhances responsive targeting. Forward observers equipped with an Android smartphone and unmanned aerial systems (UASs) use an encrypted network to input enemy targets that are transmitted and seen simultaneously, not sequentially, at all levels of command and coordination for approval. At the same time, artillery units within range of the target can select and execute technical firing solutions and service the target. Ukraine's Uber for artillery flattens and expedites approving fires and helps to achieve "any sensor, best shooter.
>
> (Bradley III 2024, 101)

Ukraine, in a whole of society effort, has changed the conduct of artillery operations in 21st-century warfare through their innovative Kropyva software. This software has reduced, or even eliminated, myriad layers of artillery fire mission assignment and allocation by providing a real-time common operating picture (COP) for both forward observers and artillery units alike, allowing the former to upload a target requirement through Kropyva's software on a battlefield with real-time connectivity thanks to Starlink, while the latter is subsequently able to service those highest priority and close-by targets in minutes. The app is named in a tradition stemming from the AFU, whereby artillery weapons systems are traditionally named after flowers and missile systems named after weather phenomena; Ukrainians now name software after plants that have a self-defence mechanism, with "Kropyva" being the Ukrainian word for a stinging-nettle plant (Lange 2023, 12).

Perhaps more importantly than its name is the fact that it was developed "by volunteers of the 'Armija SOS' initiative in support of the Armed Forces of Ukraine" (Lange 2023, 12; Army SOS 2024). According to Nico Lange's report, which draws heavily from first-hand interviews with citizens of Ukraine and soldiers of the AFU, this software app was developed for use on Android phones or tablets. Lange further notes the importance and ingenuity of the software, in that it gives the AFU formations at the lowest echelon "an up-to-date picture of the situation … Nothing runs without up-to-date data or data transfer and nothing happens without the data being immediately transferred back into the system. The app … was developed at breakneck speed after the invasion started" (Lange 2023, 12). Lange goes on to note that there are software specialists embedded in the field with the AFU that are constantly developing and improving the apps and software "during ongoing battles and always oriented towards solving practical problems that arise in war" (Lange 2023, 12). This whole of society approach has not only innovated the way in which artillery fires are conducted at all echelons of the AFU, but also has evidenced valuable lessons and best practices. These are namely improving the efficiency and effectiveness of artillery operations in war through human capital, thereby overcoming personnel and resource shortfalls and saving lives, as well as the continuous development and subsequent improvement of systems in the conduct of this war and for future wars. Also, developed and updated outside of traditional military procurement methods, Kropyva is a master-class and paradigm shift in defence procurement as well as finding synergies between military necessities and economic sector human capital and capabilities development in defence of the nation.

In summary, on artillery fires regarding Kropyva, the authors of *A Call to Action: Lessons from Ukraine for the Future Force* succinctly state:

> The functionality and responsiveness of this command and control (C2) system streamlines tactical and technical fire-direction processes to provide Ukraine a competitive advantage. Resourced outside the military procurement process and open to updates, these applications give Ukraine a technical advantage in delivering responsive fires as well as show versatility as the Armed Forces of Ukraine's current common-operational-picture platform.
>
> (Bradley III 2024, 101)

Perhaps the most important lesson learned here is that the AFU developed this software outside of the traditional military procurement process, which both enabled it to be rapidly implemented at a comparatively low cost, as well as allowing for updates and improvements without the same bureaucratic inertia that an initial procurement and subsequent updates or improvements would entail. The bottom line is that Ukraine has in many ways mitigated significant personnel and materiel shortfalls and disadvantages through the efficient and effective allocation of artillery fires resources via the Kropyva software and associated Android application that is available at all echelons and to all forces that are fighting for Ukraine against Russia's war of aggression against Ukraine.

In regards to mass of fires, Russia's war of aggression against Ukraine has shown that while PGMs have a key role to play in the conduct of 21st-century warfare, mass of fires is still a key concept for employing the king of battle, or artillery. These PGMs will continue to be used, as they were by the U.S. in the First Gulf War, to degrade or destroy critical infrastructure and capabilities of the enemy, or in other words high-value targets that provide a return on investment for such a high-cost technical weapon as a PGM. In turn, both precision and dumb artillery rounds will continue to be used to mass fires on the battlefield, primarily at the tactical level, in order to deny the enemy freedom of manoeuvre as well as in support of friendly force operations, specifically preparing the battlefield for friendly manoeuvre operations. The mass of fires will also contribute to the attrition of enemy personnel, equipment, and war materiel at both the tactical and operational levels, and in turn serves to prevent the enemy from seizing and holding key terrain, which is a critical component of success in LSCO. The bottom line is precision fires have a key role in 21st-century warfare, but Russia's war of aggression against Ukraine has shown that mass of fires will continue to be a key component at the tactical and operational levels of land warfare as well. The AFU's ability to more efficiently and effectively deliver those mass fires on valuable targets serves to not only increase the destruction of RF personnel, equipment, and war materiel, thereby decreasing RF combat power, but also reduces the logistics burden on the AFU (i.e. more efficient and effective artillery fires results in less rounds to bring to the guns and less maintenance to conduct on the guns) and likewise increases the survivability of Ukrainian artillery weapons systems by reducing the duration of their fire missions and therefore reducing exposure to counter-battery fires, or in short, allowing them to "shoot and scoot" to great effect.

In regards to the demand signal for ammunition, Russia's war of aggression against Ukraine has shown that both Ukraine's and the West's artillery ammunition estimated consumption rates in sustained LSCO were woefully underestimated and have proven insufficient in combat operations by the AFU in Russia's war of aggression against Ukraine (Zabrodskyi, Watling, Danylyuk, and Reynolds 2022, 56). This led to a corresponding underestimation on required stocks and production rates of artillery ammunition, both prior to and during the conduct of LSCO. Worth noting briefly here, this is a shortcoming of Western DIB production, specifically in regards to dumb artillery ammunition but also regarding war materiel in general, and one that needs to be solved now by all allied governments in order to provide adequate artillery ammunition and other war materiel in support of the AFU as well as in order to ensure there are sufficient stocks for any future conflict (Lange 2023, 26).

Key Takeaways and Implications

Based on the aforementioned reporting, research reports, and books from 2022 to present and the corresponding observations about the conduct of UAS and artillery operations in Russia's war of aggression against Ukraine, a few key takeaways and implications follow.

The key takeaways from the above topics is that UAS, counter-UAS, and artillery operations are critical aspects of this conflict and will be crucial to any success in 21st-century warfare. UAS and counter-UAS equipment, force-structure, and doctrine for employment is needed at all tactical echelons and even the operational level of warfighting formations. The primary function of these UAS are delivering lethal effects to enemy personnel, equipment, formations, and war materiel from tactical through strategic echelons. These UAS can also serve as intelligence collection platforms and logistics support platforms. In all cases, UAS serves to: (1) reduce the required manpower on a battlefield, serving as a way to deliver lethal munitions in order to reduce enemy resources and likewise mitigate one's own shortfalls and disadvantages in personnel and equipment, (2) provide better battlefield awareness through ISR thus reducing required reconnaissance personnel, and (3) provide sustainment to forward troops without requiring support troops and associated logistics equipment to traverse the battlefield to make such deliveries. Likewise, artillery operations increased efficiency mitigates disadvantages in personnel and equipment as well as reduces logistical burdens (i.e. a reduction in ammunition delivery and artillery gun maintenance) by ensuring more effective fires are delivered on enemy targets with the fewest possible rounds.

Another key takeaway is that there also needs to be a reduction of DIB procurement bureaucracy. As evidenced in this war with the development of both UAS platforms and artillery operations platforms that increase lethality, efficiency, and effectiveness (not to mention survivability of troops by eliminating threats) finding synergies between military necessities and economic sector capabilities and reducing any and all bureaucratic impediment between the end-user need and the developing and delivery of that capability is urgently necessary. This facilitates innovation and increases lethality of war-fighters by delivering necessary equipment and capabilities on significantly reduced timelines from conceptualisation/demand signal to operationally capable/available to the war fighter in combat operations.

The evolution of UAS that has been evidenced on the battlefields in Russia's war of aggression against Ukraine have implications for the conduct of this conflict and future wars that are paradigm shifting. Primarily, the development and integration of AI into UAS sensor and shooter decision-making. Conceptually, drones can and will be made more autonomous as they evolve in parallel to be able to carry larger and more deadly lethal munitions payloads; the end state is a short-range or long-range drone, loaded with lethal munitions, that is either able to loiter in a given area of operations or be launched on set coordinates with the ability of integrated AI to identify viable military targets and prosecute lethal munitions strikes without a human controller. The first military to fully develop this capability, which ostensibly allows for kamikaze missions on strategic targets deep into enemy rear, as well as faster kill-chain decision making at the tactical level with positively identifiable military targets, will have a significant advantage. One foreseeable problem in this integration of autonomous AI is with use of captured enemy equipment, whereby the force employing the former would have to ensure that its autonomous UAS with integrated AI platforms would not strike the latter captured enemy equipment being employed by their own friendly forces. The drone arms-race for AI-controlled lethal drones and AI controlled autonomous lethal drones, something not yet evidenced in Russia's war of aggression against Ukraine, is a critical arms-race for success in 21st-century combat operations.

Likewise, the development of software in order to provide a persistent and pervasive COP of the battlespace, facilitated through UAS possibly coupled with AI (for battlefield/target information processing, and so on) or otherwise, has significant implications for this conflict and future wars. Previously such real-time and pervasive overwatch was only available for the highest-level missions, such as small-unit formations conducting strategic level military operations or special operations forces going after high-value targets. Now, with the development of software and associated applications, almost all units down to the lowest tactical echelon have real time ISR and a COP if they so choose (i.e. infantry squads can operate with these Android devices in hand and are able to employ UAS for their own hip-pocket ISR and battlefield awareness). A further evolution with significant implication would be integrating this persistent and ubiquitous COP with AI in a manner that allows that AI to identify threats to friendly formations (or other relevant battlefield developments), and potentially even prosecute lethal strikes against those threats without input from the friendly forces on the ground that are at risk.

Conclusions

Ukraine and the West must increase their respective DIB production capacities. In order to fight and win, this conflict and future wars will require massive amounts of drones and artillery ammunition. Increasingly lethal and precise FPV drones can also mitigate personnel and resources shortfalls and disadvantages, such as artillery ammunition shortfalls, providing an innovative and often more effective and efficient strike capability than artillery. The side that innovates, employs, fails, and innovates again is likely to succeed in this conflict and in future wars. Therefore, it is crucial to find synergies between military requirements and economic capabilities, and likewise to remove all bureaucratic obstacles and inertia that would slow down this innovation, employment, failure, innovation, and success cycle. A huge part of this cycle is also the human capital available to facilitate such developments. It is crucial that the human capital of a country, both in military formations and economic sectors providing support thereto, are properly handled in order to facilitate this whole-of-society approach to fighting and winning wars, in which the speed of innovation will be critical. Finally, change in current tactics and doctrine is necessary, along with the development of anti-drone countermeasures at scale, in order to increase survivability of contemporary forces on the battlefield. Nowhere is safe any longer; extensive causalities are the reality of contemporary war with ubiquitous and lethal UAS and artillery, and likely to significantly increase with the integration of AI into these military platforms with lethal capabilities.

References

Army SOS. 2024. “Improving Ukraine’s Defense Capabilities Since 2014.” https://armysos.com.ua/

ArmyInform (@armyinformcomua). 2024. “The “Bulava” Unit Has Developed a New #FPVdrone Called “Queen of Hornets,” Capable of Firing a Handheld Anti-Tank Grenade Launcher to Support Infantry in Assaults and Destroy Enemy Equipment, Enhancing Soldier Safety on the Battlefield.” X (formerly Twitter), September 21. https://x.com/armyinformcomua/status/1837454888512475353

Bowden, Mark. 2022. “The Tiny and Nightmarishly Efficient Future of Drone Warfare: Russia’s war on Ukraine has given us just a peek of the world to come.” *The Atlantic*, November 22. www.theatlantic.com/technology/archive/2022/11/russia-ukraine-war-dronesfuture-of-warfare/672241/

Bradley III, John “Jay” B. 2024. “Fires.” In *A Call to Action: Lessons from Ukraine for the Future Force*, edited by John A. Nagl and Katie Crombie. US Army War College Press.

Freedman, Lawrence. 2023. *Modern Warfare: Lessons from Ukraine*. Penguin Books.

General Headquarters Armed Forces of Ukraine (@GeneralStaffUA). 2024. “Загальні бойові втрати противника з 24.02.22 по 30.09.24 орієнтовно склали / The Estimated Total Combat Losses of the Enemy from 24.02.22 to 30.09.24.” X (formerly Twitter), September 30. https://x.com/GeneralStaffUA/status/1840247153219817576

Hammes, Thomas X. 2023. “Game Changers: Implications of the Russo-Ukraine War for the Future of Ground Warfare.” *Atlantic Council Issue Brief*. www.atlanticcouncil.org/in-depth-research-reports/issue-brief/game-changers-implications-of-the-russo-ukraine-war-for-the-future-of-ground-warfare/

Holbrook, Matthew S. 2024. “Protection: Electronic, Air, Civilian, and Infrastructure.” In *A Call to Action: Lessons from Ukraine for the Future Force*, edited by John A. Nagl and Katie Crombie. US Army War College Press.

Kirichenko, David (@DVKirichenko). 2024. “I Spoke with a Tank Crew Fighting in the Battle for Toretsk. Here Is What They Told Me About How Tank Warfare. ‘The Era of the Cautious Tank.’ ” X (formerly Twitter), September 16. https://x.com/dvkirichenko/status/1835701547281600696

Lange, Nico. 2023. “How to Beat Russia: What Armed Forces in NATO Should Learn from Ukraine’s homeland defense.” *GLOBSEC*. www.globsec.org/sites/default/files/202302/How%20to%20beat%20Russia%20by%20Nico%20Lange%20v7%20web.pdf

MacKay, Michael (@mhmck). 2024. “The Protection the United States Gives to the Russian Terrorist State Is Driving Innovation in Ukraine. Defenders Are Developing Cruise Missiles with the Range and Punch of the Weapons the U.S. Refuses to Provide. Ukraine’s ‘Palianytsia’ Jet Drone Is Already Striking the Enemy.” X (formerly Twitter), August 24. https://x.com/mhmck/status/1827399662543740934

NATO Partnership for Peace Consortium (PfPC). 2023. “Russia’s War Against Ukraine Lessons Learned Curriculum Guide.” *NATO Headquarters*. www.nato.int/cps/en/natohq/topics_221175.htm

NEXTA (@nexta_tv). 2024. "Tests of Ukrainian FPV drone with RPG-18 grenade launcher." X (formerly Twitter), September 10. https://x.com/nexta_tv/status/1833551296919384403

Trofimov, Yaroslav. 2024. *Our Enemies Will Vanish: The Russian Invasion and Ukraine's War of Independence*. Penguin Press.

Urbancik, Johanna and Denys Glushko. 2024. "Wild Hornet Attacks: How Ukraine's Drones Are Making Their Mark on the Frontline." *Euronews*, September 26. www.euronews.com/next/2024/09/26/wild-hornets-ukraine-drones

Watling, Jack, Oleksandr V Danylyuk, and Nick Reynolds. 2024. "Preliminary Lessons from Ukraine's Offensive Operations, 2022-23." *Royal United Services Institute*. https://static.rusi.org/lessons-learned-ukraine-offensive-2022-23.pdf

Wild Hornets (@wilendhornets). 2024a. "We Created a New Kind of Drone to Shoot Down Russian Spies. In Two Months, More Than 100 Scouts Were Shot Down by Such Drones. In the Video You Can See How It Happens." X (formerly Twitter), August 30. https://x.com/wilendhornets/status/1829620332451270884

Wild Hornets (@wilendhornets). 2024b. "Our Military Continues to Develop the Hornet Queen Equipped with Automatic Weapons This Time, the First Combat Deployment Was Carried Out—Targeting a Position With Russian Forces." X (formerly Twitter), September 4. https://x.com/wilendhornets/status/1832810322735890925

Wild Hornets (@wilendhornets). 2024c. "Not Just a Wild Hornet, but a Drone-Dragon." X (formerly Twitter), September 8. https://x.com/wilendhornets/status/1832810322735890925

Wild Hornets (@wilendhornets). 2024d. "Successful Test of the World's First Rocket Launcher Drone." X (formerly Twitter), September 13. https://x.com/wilendhornets/status/1834583491406700683

Zabrodskyi, Mykhaylo, Jack Watling, Oleksandr V Danylyuk, and Nick Reynolds. 2022. "Preliminary Lessons in Conventional Warfighting from Russia's Invasion of Ukraine: February–July 2022." *Royal United Services Institute*. www.rusi.org/explore-our-research/publications/special-resources/preliminary-lessons-conventional-warfighting-russias-invasion-ukraine-february-july-2022

12 Conclusions

Tracey German, Fotios Moustakis, and Andrew N. Liaropoulos

Key insights

As warfare evolves, the interplay of military strategy and innovation has become more critical than ever. *The Co-evolution of Technology and Warfare: Reshaping the Battlefield* explores this transformation, analysing the profound impact of emerging technologies on modern conflict. The following chapter summaries conclude the book, setting out key insights into the implications of these advancements for contemporary warfare. By examining global military trends and recent conflicts, these final reflections highlight both the opportunities and challenges of modern warfare, offering a comprehensive outlook on the future of security and defence in an uncertain global system.

The chapter by Jack Sharpe on the intersection of artificial intelligence (AI) and cyber warfare, focused on its transformative impact on global security. It discussed the strategic advantages of AI in cyber operations, such as enhancing attack speed and improving intelligence gathering. The chapter provided an in-depth analysis of AI-driven cyber threats, particularly in critical infrastructure attacks, espionage, disinformation campaigns, and the increasing use of AI-powered malware. Russia's cyber operations, including disinformation tactics and AI-enhanced hacking strategies, were analysed within the context of the ongoing conflict in Ukraine, illustrating how AI-driven cyber warfare is shaping modern conflicts. Future trends in AI-enhanced cyber warfare, including quantum computing and AI-powered cyber-physical attacks, were also discussed, emphasising the urgency of developing proactive counter-measures. Recommendations for policymakers, military leaders, and cybersecurity professionals focus on investing in AI-driven cybersecurity solutions, strengthening global collaboration, and maintaining human oversight in AI decision-making. Sharpe concludes that AI is revolutionising cyber warfare, calling for the urgent adaptation of security strategies to manage both opportunities and risks effectively.

Andrew Liaropoulos explored how AI transforms influence campaigns by enabling the automated creation and dissemination of content. While such operations have historically been used as a military and political strategy, AI-powered tools, such as chatbots, deepfake technologies, and generative language models (LLMs),

DOI: 10.4324/9781003520160-12

have enhanced their effectiveness, making influence campaigns more cost-effective and personal. The chapter examined how AI amplifies influence by automating content generation, reducing human labour, and increasing the reach of state and non-state actors, including militaries, political organisations, and private firms. It highlights how AI expands actors, allowing states, non-state groups, and private firms to wage influence campaigns; enhances behaviours, reducing costs and improving targeted disinformation; and generates deceptive content, including deepfake videos and automated social media manipulation. On the other hand, Liaropoulos identifies limitations, including bias in AI models, detection efforts, and challenges in gaining audience attention. The chapter concludes that AI's role in IOs must be countered with regulation, media literacy, and AI-powered detection tools to prevent societal polarisation and distrust.

The chapter by Chris Lavers explored the growing threat of commercially available first-person view (FPV) unmanned aerial vehicles (UAVs) modified for asymmetrical warfare, particularly by non-state actors such as Al-Qaeda, ISIS, and Hezbollah. It discussed how commercially available drones have been weaponised to conduct surveillance, assassinations, and infrastructure attacks, with examples from conflicts in Syria, Iraq, and Ukraine. UAVs have evolved from military tools to widespread commercial platforms, blurring the lines between civilian and military use. Advances in drone technology have made them highly attractive to terrorist groups due to their affordability, ease of modification, and stealth capabilities. Modified UAVs offer insurgents significant tactical benefits, including the ability to swarm, evade detection, and deliver precision strikes. Cheap, commercially available drones can be adapted for various hostile applications, from reconnaissance to carrying explosives. The proliferation of modified drones poses also security risks to critical infrastructure and military installations. The chapter highlights the urgent need for policy interventions and regulatory frameworks to mitigate UAV threats. Measures such as radar enhancements, airspace restrictions, and drone-detection systems must evolve to counteract these emerging dangers.

Markos Trichas and Matthew Mowthorpe explored the evolving landscape of space warfare, by focusing on the increasing militarisation of space by Russia and China. During the Cold War, space was largely treated as off-limits for military aggression, culminating in the 1967 Outer Space Treaty. However, in recent decades, space has become an operational military domain, with nations actively developing counterspace capabilities. The chapter highlights the strategic importance of space assets in modern warfare. Satellites support communications, navigation, intelligence, and precision strikes, making them critical to military operations. The conflict in Ukraine demonstrated this when Russia launched a cyber-attack on Viasat, disrupting communications, while Ukraine utilised Starlink to maintain battlefield connectivity. Russia's counterspace program has advanced significantly under Vladimir Putin, focusing on electronic warfare (EW), co-orbital anti-satellite (ASAT) systems, and direct ascent weapons. China has similarly prioritised space superiority. Its counterspace program includes kinetic ASAT tests, co-orbital satellites with robotic arms capable of grappling objects, and directed-energy weapons for blinding or damaging satellites. China has demonstrated the ability to manoeuvre satellites covertly and potentially interfere with adversary space assets. The chapter concludes that space is now a contested battlefield, necessitating rapid and innovative defence strategies. A multi-dimensional space-based architecture, integrating manoeuvrable satellites and decoys, is proposed to enhance resilience against ASAT threats. Both Russia and China's developments underscore the urgent need for international coordination to safeguard space assets from escalating hostilities.

The chapter by Tracey German argues that hypersonic weapons, capable of speeds above Mach 5, are reshaping military strategy with their speed, precision, and manoeuvrability, challenging existing defence systems. Used in combat for the first time by Russia in Ukraine (2022), their impact has been underwhelming, with limited strategic gains. There are two main types: hypersonic cruise missiles (HCMs) and hypersonic glide vehicles (HGVs). Their unpredictable flight paths make them difficult to intercept, raising concerns about deterrence and strategic stability. While hypersonics complicate global power balances and defence planning, they remain expensive and scarce. Their real impact is more strategic than battlefield-defining, challenging traditional deterrence models but not yet revolutionising warfare. In conclusion, the chapter stresses that hypersonics are a significant technological advancement but have yet to prove themselves as a decisive game-changer in warfare. Their primary impact lies in their strategic implications rather than battlefield effectiveness.

The next chapter by Sidharth Kaushal explored how globalisation has reshaped naval strategy, particularly in the context of the US-China maritime rivalry. A key argument is that sea control is shifting from distant blockades to operations in the littoral regions. The complexity of modern shipping – flag changes, multiple ownership structures, and transactions at sea – makes traditional interdiction challenging. This benefits China, which, due to its geographic proximity and large coast guard, can exert greater influence over regional trade than the US. However, the US is adapting its strategy. While China's dependence on trade and shipbuilding dominance gives her advantages, the US retains strategic offsets such as advanced submarines and long-range missile capabilities. Unlike past sea powers, the US may function more like Cold War-era continental states (e.g., the USSR), focusing on sea denial rather than sea control. Another trend discussed is the reintegration of government into sea control. The US and China are increasingly leveraging commercial tools like maritime insurance, undersea cables, and shipping infrastructure to exert influence. Ultimately, the chapter argues that the US must rethink its naval strategy, learning from past continental challengers rather than traditional seapowers. The competition in the South China Sea will be shaped by regionalised trade, economic statecraft, and the ability to control commerce near contested waters. While China has advantages in shipbuilding and trade reliance, the US holds asymmetrical strengths, making future naval competition highly dynamic.

James Henry Bergeron examined the evolving landscape of maritime strategy in the context of the Second Revolution in Military Affairs (2RMA). Historically, post-Cold War maritime strategies prioritised power projection and crisis response over traditional naval defence. However, shifting geopolitical realities, including Russia's increasing aggressiveness, China's naval expansion, and the proliferation of autonomous military technology, necessitate a strategic recalibration. The chapter outlined key factors driving this transformation. The 2022 Russian invasion of Ukraine demonstrated how modern, cost-effective technologies like drones and autonomous systems can challenge established naval forces. Ukraine's innovative use of land-based anti-ship missiles and maritime drones forced the Russian Black Sea Fleet into retreat, proving that smaller nations can wield asymmetric capabilities effectively. Similarly, Houthi rebels have leveraged drones and missiles to disrupt global trade routes in the Red Sea, showcasing how non-state actors can challenge naval supremacy. The 2RMA is marked by the integration of AI, cyber warfare, space-based assets, and hypersonic weapons, reshaping naval tactics and force structures. Debates persist on whether to invest in conventional warships or prioritise autonomous platforms, electronic warfare, and counter-drone technologies. The chapter stressed that strategically, NATO and Western powers must leverage their maritime superiority to deter adversaries like Russia and China. This involves securing sea lines of communication, enhancing undersea surveillance, and preparing for potential maritime blockades. The future of naval warfare will require rapid adaptation, strong industrial support, and sustained political commitment to maintain maritime dominance in an era of disruptive technological change.

The challenges that hybrid and cyber threats pose to critical infrastructure (CI) was explored by Konstantinos Tsetsos. These threats combine disinformation, cyberattacks, economic coercion, and social engineering, operating in the ambiguous grey zone between peace and conflict. Fifth-generation warfare tactics, including AI-driven cyberattacks and psychological manipulation, exploit attribution challenges, making defence and response challenging. The chapter outlined the evolution of warfare from traditional military conflicts to modern hybrid threats, where non-kinetic attacks such as propaganda, cyber sabotage, and economic pressure are used to destabilise nations. Cyberattacks on CI, such as power grids, healthcare systems, and financial services, have increased significantly, with state and non-state actors exploiting vulnerabilities in interconnected systems. The effects of such attacks can disrupt entire societies, making robust security measures critical. The chapter emphasised the importance of a resilience-based security culture, advocating for proactive threat detection, early warning mechanisms, and policy-driven cybersecurity investments. Governments, organisations, and communities must collaborate to transform security from a reactive stance into a foundational element of societal stability, ensuring continuity of essential services amid evolving cyber and hybrid threats.

The evolving landscape of warfare, marked by rapid technological advancements, requires a fundamental shift in military leadership and training. The chapter by Fotios Moustakis explored how AI, robotics, cyber warfare, and autonomous systems are reshaping military strategies and challenging traditional Western supremacy. These technologies enable adversaries to act swiftly and creatively, often outpacing democratic nations bound by bureaucratic constraints. To maintain their strategic edge, Western military organisations must adopt transformational leadership, which emphasises adaptability, innovation, and long-term vision over traditional transactional models. Effective leadership in high-tech warfare demands not only technical competence but also an understanding of the human dimension, motivating troops, fostering innovation, and aligning technological advancements with operational objectives. The Ukrainian resistance against Russia illustrates the power of adaptive leadership combined with modern technology. Military leadership today must navigate complex environments that shift rapidly between peace and war. Training must emphasise agility, versatility, and continuous education. Leaders must not only execute tactical operations but also ensure the well-being of their personnel and maintain institutional integrity. Transformational leadership, which fosters creativity, problem-solving, and adaptability, is increasingly recognised as vital in military organisations. The chapter concludes that as new technologies redefine warfare, military training should incorporate academic, industrial, and practical simulations to enhance decision-making and technological integration.

Finally, the chapter by Danny Love examined the tactical innovations undertaken by the Armed Forces of Ukraine (AFU) in response to Russia's full-scale invasion. The analysis focused on two key areas: unmanned aerial systems (UAS) employment and artillery operations. These innovations highlight Ukraine's ability to adapt and leverage technology against a larger, better-resourced adversary. Ukraine has pioneered the large-scale use of cheap, disposable drones to deliver lethal strikes. Unlike traditional UAS platforms used for reconnaissance, Ukraine has modified quad-copter drones to deploy grenades, thermite, and even anti-tank weapons. Ukraine has also revolutionised artillery warfare through software-driven targeting and fire coordination. The Kropyva system, an Android-based application, enables rapid artillery strikes by linking forward observers, drones, and artillery units in real time. The key highlights from this final chapter demonstrate that drones are revolutionising modern warfare by enhancing intelligence, surveillance, and reconnaissance (ISR), enabling precise targeted strikes, and supporting logistics operations. Meanwhile, artillery remains a critical component of combat, with software-driven coordination significantly improving its efficiency. The integration of AI in UAS is expected to play a decisive role in future conflicts. These advancements not only demonstrate Ukraine's adaptability but also provide valuable lessons for Western militaries preparing for the complexities of 21st-century warfare.

Future projections

The future of warfare is a topic of intense debate, shaped by technological advancements, geopolitical shifts, and evolving military doctrines. As states prepare for future conflicts that may differ significantly from those of the past, several critical debates emerge, encompassing AI, cyber warfare, autonomous weapons, space militarisation, hybrid warfare, ethical considerations, and the impact of emerging technologies on global security dynamics. This book analysed some of these debates, but there are also other trends that shape the future face of warfare.

Looking ahead, modern militaries are expected to dramatically expand AI's role in warfare. As technology advances, military AI will likely evolve from its current peripheral and supportive roles to playing a more central role in combat and strategic decision-making. Conflict zones have already seen an increasing adoption of AI for intelligence gathering, with platforms like Palantir's AI responsible for most targeting Ukraine operations on the battlefield against Russian forces. The use of AI-driven intelligence highlights its critical role in contemporary warfare. The speed of AI processing and advanced algorithms are essential, enabling commanders to make faster and more well-informed decisions. AI can rapidly analyse reconnaissance data and translate it into actionable insights within seconds, significantly enhancing battlefield efficiency.

Projections of future warfare, particularly in Western military assessments and visualisations, suggest that the integration of AI-driven judgment with senior officers' decision-making mechanisms will become even more intertwined. This human–AI collaboration will leverage the strengths of both AI's speed and data-processing capabilities combined with human intuition and legal or ethical considerations. Despite AI's growing capabilities, the prevailing assumption remains that AI will serve as a tool for decision-making rather than being an autonomous decision-maker.

AI-powered predictive analytics, which uses data to forecast future trends and events, is also expected to play a key role in military logistics. Currently utilised in the commercial sector for demand prediction and delivery optimisation, predictive analytics will be adapted to ensure troops receive and maintain necessary supplies at the right time. Commanders in the field will use AI to assess supply adequacy and anticipate potential disruptions, enabling logistics teams to adjust supply lines proactively. Just as Lt. General Pagonis streamlined logistics under a unified command during the first Gulf War, AI will enhance centralised decision-making by aggregating vast amounts of real-time data to optimise supply chains and predict future needs.[1] Additionally, the flexibility of Pagonis' decentralised execution – where local commanders retained autonomy – parallels AI-driven logistics models, which allow real-time adjustments based on evolving battlefield conditions, weather changes, or enemy movements.

In the near future, battlefields will increasingly be dominated by autonomous and semi-autonomous weapon systems, ranging from smart missiles and drones to robotic tanks and autonomous submarines. These advancements will fundamentally reshape warfare, making AI-driven systems the norm. While human-crewed assets will remain, AI-assisted platforms like the Kratos XQ-58A Valkyrie, designed as a "loyal wingman" for scouting, defence and enemy engagement, will become standard. Capable of operating within drone swarms without direct human piloting, these AI-enhanced systems will play a crucial role in modern combat.

Similarly, land and naval forces will deploy autonomous weapon systems to execute attacks and guard defensive positions, with military commanders setting mission objectives. AI will also become an essential tool for real-time battlefield simulations. While AI may not predict the overall outcome of a war, it will run countless simulations during battle, continuously updating them based on shifting battlefield conditions and offering revised tactical recommendations.

Equally important is the integration of AI-enabled systems into the Joint All-Domain Command and Control (JADC2) concept, developed by the U.S. Department of Defense. JADC2 aims to operate across all levels and phases of warfare, spanning multiple domains and coalition partners, to maintain an information advantage at the speed of relevance.

While these advancements will significantly reshape warfare, maintaining meaningful human control over AI-driven weapons and decisions remains a fundamental principle. Ensuring meaningful human control over AI-driven weapon systems and decision-making processes remains a priority. Maintaining this control can be challenging, particularly as AI operates in fractions of a second. However, the guiding principle within Western armed forces is, and should continue to be, that militaries maintain meaningful human oversight over weapons and critical decisions.

It could be argued that the future battlefield will be dominated by software. Algorithms might determine the success of military missions more than platforms, as a result of disruptive developments and the introduction of faster communication networks and system-of-systems defence solutions. In the coming two decades, major global powers will have a fully established space force resulting in new opportunities but also threats. The consequences of space conflicts could be catastrophic, as damaging even a few satellites could disrupt global communications, navigation, and financial systems. Such a scenario will call for the establishment of new concepts of operations, regulatory frameworks, and international agreements. Another trend that we have to take into account is the merger of brain–computer interfaces and augmented/virtual reality on the battlefield. Future high-tech helmets and smart glasses will optimise a soldier's situational awareness, shape perception, compress time, accelerate decision-making, and once again promise to lift the fog of war. There may well be further integration and cooperation between humans and machines in future conflict. The development of hybrid human–machine squadrons raises major concerns regarding the ethical and regulatory framework of future warfare. Adding to that, key advances in biotechnology, synthetic biology, and brain–computer interfaces will enhance the cognitive abilities of soldiers and challenge our understating of warfare. The future commander will utilise quantum sensing and quantum navigation, making the promise of real-time mapping a reality.

The future of warfare is uncertain. What is certain though is that technology defines warfare, but does not win or lose wars on its own. History is a reminder that technology is one of the many factors that shape the conduct of warfare. There is no doubt that technology will alter how future militaries plan for and fight wars, but the decision-makers cannot lose sight of the human and political domains of war regardless of the technological advances.

Note

1 William G. Pagonis and Jeffrey L. Cruikshank, *Moving Mountains: Lessons in Leadership and Logistics from the Gulf War* (Brighton, MA: Harvard Business Review Press, 1992).

Index

FSC
www.fsc.org
MIX
Papier | Fördert gute Waldnutzung
FSC® C083411

Kolibri 360 GmbH
Ferdinand-Jühlke-Straße 7
99095 Erfurt, Deutschland
produktsicherheit@kolibri360.de